Mathematics for Industry

Volume 4

For further volumes:
http://www.springer.com/series/13254

Aims & Scope

The meaning of "Mathematics for Industry" (sometimes abbreviated as MI or MfI) is different from that of "Mathematics in Industry" (or of "Industrial Mathematics"). The latter is restrictive: it tends to be identified with the actual mathematics that specifically arises in the daily management and operation of manufacturing. The former, however, denotes a new research field in mathematics that may serve as a foundation for creating future technologies. This concept was born from the integration and reorganization of pure and applied mathematics in the present day into a fluid and versatile form capable of stimulating awareness of the importance of mathematics in industry, as well as responding to the needs of industrial technologies. The history of this integration and reorganization indicates that this basic idea will someday find increasing utility. Mathematics can be a key technology in modern society.

The series aims to promote this trend by (1) providing comprehensive content on applications of mathematics, especially to industry technologies via various types of scientific research, (2) introducing basic, useful, necessary and crucial knowledge for several applications through concrete subjects, and (3) introducing new research results and developments for applications of mathematics in the real world. These points may provide the basis for opening a new mathematics-oriented technological world and even new research fields of mathematics.

Ken Anjyo
Editor

Mathematical Progress in Expressive Image Synthesis I

Extended and Selected Results from the Symposium MEIS2013

Springer

Editor
Ken Anjyo
OLM Digital, Inc.
Tokyo
Japan

ISSN 2198-350X ISSN 2198-3518 (electronic)
ISBN 978-4-431-55006-8 ISBN 978-4-431-55007-5 (eBook)
DOI 10.1007/978-4-431-55007-5
Springer Tokyo Heidelberg New York Dordrecht London

Library of Congress Control Number: 2014939620

Printed on acid-free paper

Springer is part of Springer Science+Business Media (www.springer.com)

Preface

Over the last decade, computer graphics (CG) modeling has developed considerably, bringing greater fidelity and interaction in many applications. However, subjects such as fluids and virtual humans continue to pose CG challenges. The symposium "Mathematical Progress in Expressive Image Synthesis" (MEIS2013) was held in Fukuoka, Japan, in October 21–23, 2013, as a unique venue where mathematicians, CG researchers, and those who work in industry came together to investigate these difficult subjects. Moreover, the symposium was intended to trigger novel research directions such as exploring the mathematics of visual perception.

This volume presents the papers selected from the MEIS2013 proceedings, which was originally issued as *MI Lecture Notes* Vol. 50, Kyushu University, 2013. The book comprises five parts in order to organize the papers for mathematical scientists, graphics researchers, and industry programmers as well. Part I presents several mathematical frameworks and approaches relevant to CG and vision issues on a widespread scale. Each theme of Part II–IV is a more specific 2D/3D CG issue regarding photorealistic rendering, texture and sound synthesis, visual simulation of fluids, surface deformation/editing, and character locomotion. Part V presents papers on how to utilize 2D image databases for practical applications. The underlying mathematical subjects involve discrete differential geometry, Lie theory, computational fluid dynamics, function interpolation, and learning theory. We therefore hope the readers will find themselves deeply inspired through the harmony of mathematics and graphics research displayed in this volume.

Tokyo, February 2014 Ken Anjyo

Acknowledgments

The MEIS2013 Organizing Committee is very grateful to the Institute of Mathematics for Industry (IMI), Kyushu University, for the sponsorship of the Symposium MEIS2013. Our deepest thanks also go to the Japan Science and Technology Agency (JST), Mathematics Program: Alliance for Breakthrough between Mathematics and Sciences (ABMS), for their financial support.

The editor would also like to acknowledge the following people in creating this volume: Mathieu Desbrun (CalTech) for his great help in the paper selection process; and J. P. Lewis (Victoria University) and Makoto Okabe (The University of Electro-Communications) for their keen reviews. The editor extends his deepest thanks to Ayumi Kimura and Sampei Hirose (OLM Digital) for their hard work on the conference management and the production of this volume.

The MEIS2013 Organizing Committee

Ken Anjyo, OLM Digital, Inc.
Hiroyuki Ochiai, Institute of Mathematics for Industry, Kyushu University
Yoshinori Dobashi, Hokkaido University
Yoshihiro Mizoguchi, Institute of Mathematics for Industry, Kyushu University
Shizuo Kaji, Yamaguchi University

Contents

Part I
Mathematical Approaches to Computer Graphics and Vision

The Power of Orthogonal Duals (Invited Talk)

Mathieu Desbrun and Fernando de Goes

Keywords Orthogonal dual · Blue noise · Masonry structure · Discrete exterior calculus

Triangle meshes have found widespread acceptance in computer graphics as a simple, convenient, and versatile representation of surfaces. In particular, computing on such simplicial meshes is a workhorse in a variety of graphics applications. In this context, mesh duals (tied to Poincaré duality and extending the well known relationship between Delaunay triangulations and Voronoi diagrams) are often useful, be it for physical simulation of fluids [4] or parameterization [7]. However, the precise embedding of a dual diagram with respect to its triangulation (i.e., the placement of dual vertices) has mostly remained a matter of taste or a numerical after-thought, and barycentric versus circumcentric duals are often the only options chosen in practice. In this talk we discuss the notion of *orthogonal dual diagrams*, and show through a series of recent works that exploring the full space of orthogonal dual diagrams to a given simplicial complex is not only powerful and numerically beneficial, but it also reveals (using tools from algebraic topology and computational geometry) discrete analogs to continuous properties.

Starting from a (primal) triangle mesh defined as a simplicial complex (i.e., a piecewise linear approximation of a discrete orientable manifold surface of any topology in $\mathbb{R}^3$, with or without boundary), a family of intrinsic dual diagrams [9] can be constructed through the addition of a weight per vertex [5]. The resulting diagrams can be intuitively understood as displacements of the canonical (Euclidean)

M. Desbrun (✉) · F. de Goes
The Applied Geometry Lab, California Institute of Technology, 1200 E. California Boulevard, Pasadena, CA 91125, USA
e-mail: mathieu@cms.caltech.edu

F. de Goes
e-mail: fdegoes@cms.caltech.edu

K. Anjyo (ed.), *Mathematical Progress in Expressive Image Synthesis I*, Mathematics for Industry 4, DOI: 10.1007/978-4-431-55007-5_1,

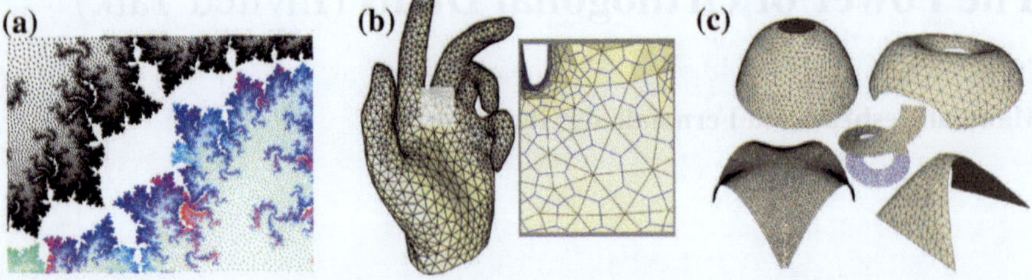

Fig. 1 Orthogonal dual diagrams to primal simplicial meshes have recently been shown key in a wide variety of applications in geometry processing and graphics, for flat & curved domains of arbitrary topology. **a** Blue noise sampling. **b** Well-centered meshes. **c** Self-supporting structures

circumentric dual along the gradient of the function defined by the weights, resulting in intrinsically straight dual edges that remain orthogonal to primal edges due to the curl-free nature of any gradient field [3]. For surfaces of non-trivial genus, there are additional displacement fields that are curl-free but are not gradients: they correspond to the so-called harmonic 1-forms (of dimension β_1, the first Betty number of the surface). Therefore, the total space of orthogonal duals is, accounting for the gauge of the gradient, of dimension $\beta_1 + V - 1$ where V denotes the number of vertices in the primal mesh. Note that once a dual diagram is defined through these coordinates, close formulae for the signed measures of the dual elements (dual lengths, dual cell areas) are available (Fig. 1).

This simple definition of orthogonal duals is surprisingly versatile in that it offers efficient and foundational solutions to numerous applications, including:

- *Blue noise sampling:* Coined by Ulichney [11], the term *blue noise* refers to an even, isotropic, yet unstructured distribution of points in (typically 2D) Euclidean space. Blue noise was first recognized as crucial in dithering of images since it captures the intensity of an image through its local point density, without introducing artificial structures of its own. It rapidly became prevalent in various scientific fields, especially in computer graphics, where its isotropic properties lead to high-quality sampling of multidimensional signals, and its absence of structure prevents aliasing. However, the generation of high-grade blue-noise importance sampling remains numerically challenging.
  Our orthogonal duals offer a convenient solution to this common requirement. By writing the density requirement as constraints on dual cell areas of the diagram and using optimal transport to formally characterize the isotropy of a point distribution, one ends up with an efficient optimization technique of point distributions via a constrained minimization in the space of orthogonal dual diagrams. In this application, the weights are crucial degrees of freedom to exactly enforce adapted sampling, rendering the formulation not only well-posed but efficient as well. In practice, the resulting blue noise point distributions outachieve previous methods based on both spectral and spatial analyses [1].

- *Self-supporting structures:* Masonry structures are arrangements of material blocks, such as bricks or stones, that support their own weight. Constructing curved vaults or domes with compression-only structures of blocks, further prevented from slipping through friction and/or mortar, has been practiced since antiquity. It is therefore no surprise that form finding and stability analysis of self-supporting structures have been an active area of research for years. In particular, it has been shown that equilibrium of a masonry structure is ensured if there exists an inner *thrust* surface which forms a compressive membrane resisting the external loads [6]. Discretizing the continuous balance equations relating the stress field on the thrust surface to the loads has been, however, an open problem for years with no fully satisfactory solution.
  Our primal-dual structures offer, here again, an unexpected approach to this problem: a finite-dimensional formulation of the compressive stress field of these self-supporting membranes represented a triangle meshes can be rigorously derived through homogenization; moreover, equilibrium is guaranteed if (and only if) the dual planar graph induced by vertex weights forms an orthogonal dual diagram—corresponding to the force network at play within the membrane. Therefore, our full characterization of orthogonal duals formally provides discrete (and exact) analogs of continuous properties; in fact, the weight themselves correspond in this context to the Airy stress function, a staple of static continuum mechanics. One can thus derive computational form-finding tools to alter a reference shape into a free standing simplicial structure, which turns out to improve upon previous work in terms of efficiency, accuracy, and scalability [2].
- *Meshing:* Being able not only to choose primal vertex positions, but also a dual diagram, opens up a series of possibilities in the context of meshing, i.e., turning a 2D or 3D domain into a(n often simplicial) complex. A majority of meshing approaches restrict the space of valid meshes to Delaunay triangulations because they are abundantly vetted by theoretical guarantees; in practice, however, Delaunay conditions are often too restrictive to be valuable. Moreover, it is extremely difficult in practice to construct "self-centered" Delaunay triangulations [10] for which each circumcenter lies inside its associated tetrahedron: failure to satisfy this property locally can lead to numerical degeneracies. Recent methods attempting to optimize meshes to avoid this issue remain impractical for complex domains [12]. With the added flexibility offered by weights, one can much more easily optimize a mesh to become well-centered by finding the weight assignment that results in dual vertices closest to each triangle's barycenter. Moreover, as our primal-dual structures are compatible with Discrete Exterior Calculus (DEC, a finite-dimensional calculus inspired by Cartan's exterior calculus), one can also optimize meshes to make a particular discrete operator (e.g., the commonly-used Laplacian) both better conditioned and with smaller error bounds [8].

Orthogonal duals have also a number of connections to other research fields, such as circle packing and discrete conformal structures, which indicates that more results are likely to come out in the next few years.

Acknowledgments Pooran Memari, Patrick Mullen, Houman Owhadi, and Pierre Alliez have significantly contributed to various parts of this work.

References

1. de Goes F, Breeden K, Ostromoukhov V, Desbrun M (2012) Blue noise through optimal transport. ACM Trans Graph (SIGGRAPH Asia) 31(4):171:1–171:11.
2. de Goes F, Alliez P, Owhadi H, Desbrun M (2013) On the equilibrium of simplicial masonry structures. ACM Trans Graph (SIGGRAPH) 32(4):93:1–93:10.
3. de Goes F, Mullen P, Memari P, Desbrun M (2013) Weighted triangulations for geometry processing. ACM Trans Graph 31:103:1–103:12.
4. Elcott S, Tong Y, Kanso E, Schröder P, Desbrun MP (2007) Stable, circulation-preserving, simplicial fluids. ACM Trans Graph 26(1):140–164
5. Glickenstein D (2005) Geometric triangulations and discrete Laplacians on manifolds. Preprint at arXiv.org:math/0508188.
6. Heyman J (1966) The stone skeleton. Int J Solids Struct 2(2):249–279
7. Mercat C (2001) Discrete Riemann surfaces and Ising model. Comm Math Phys 218:177–216
8. Mullen P, Memari P, de Goes F, Desbrun M (2011) HOT: Hodge-optimized triangulations. ACM Trans Graph (SIGGRAPH) 30(4):103:1–103:12.
9. Munkres JR (1984) Elements of algebraic topology. Addison-Wesley, Boston
10. Rajan VT (1994) Optimality of Delaunay triangulation in riptsize $\mathbb{R}^d$. Disc Comp Geo 12(1):189–202
11. Ulichney RA (1987) Digital halftoning. MIT Press, New York
12. Evan VanderZee, Hirani Anil N, Damrong Guoy, Edgar Ramos (2010) Well-centered triangulation. SIAM Journal on Scientific Computing 31(6):4497–4523

Mathematical Models of Visual Information Processing in the Human Brain and Applications to Visual Illusions and Image Processing

Hitoshi Arai

Abstract This chapter is a survey of a series of joint works of Shinobu Arai and me. The main purpose of our study is to construct mathematical models of visual information processing in the brain, and to give applications to study of visual illusions and image processing. In this chapter I will describe a part of our simulations of some visual illusions, creations of the so-called "fuyuu illusions", and an application to image processing.

Keywords Framelet · Pinwheel framelet · Image processing · Visual illusion

1 Introduction

On the past few decades, several studies have been made on mathematical models of visual information processing in the human brain. Our new models are constructed by using simple pinwheel framelets ([5]) and pinwheel framelets ([7]), which are new classes of the so-called framelets. The general scheme of framelets was established by Daubechies et al. [9]. Our simple pinwheel framelets and pinwheel framelets are new framelets appropriate to study of vision. Before going to the main body of this chapter, I will review (simple) pinwheel framelets.

H. Arai (✉)
Graduate School of Mathematical Sciences, The University of Tokyo/JST CREST, 3-8-1 Komaba, Meguro-ku, Tokyo 153-8914, Japan
e-mail: h-arai@ms.u-tokyo.ac.jp

K. Anjyo (ed.), *Mathematical Progress in Expressive Image Synthesis I*, Mathematics for Industry 4, DOI: 10.1007/978-4-431-55007-5_2,

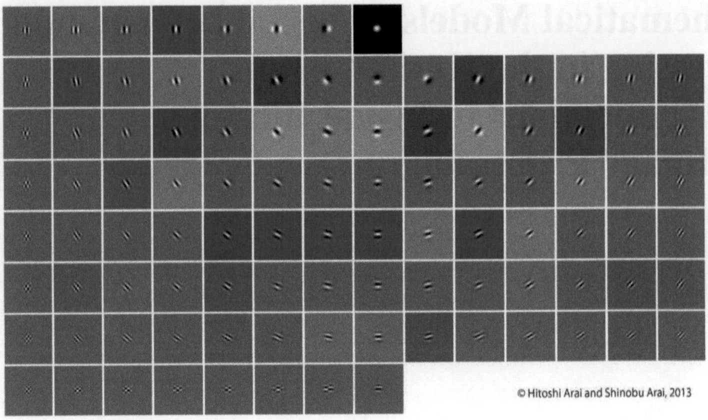

Fig. 1 Maximal overlap version of the pinwheel framelet of degree 7 (level 2)

2 Pinwheel Framelets

Some filters have been well known as models of simple cells in V1. For example, Gabor filters, DOG filters and ODOG filters. Recently R. Young pointed out that Gaussian derivatives are more appropriate than these filters (see Young [13]). Adopting his theory, Escalante-Ramírez and Silván-Cárdenas [10] proposed a mutliresolution model which consists of Gaussian derivatives. However Gaussian derivatives are not compactly supported: in other words, filters have infinite length in nature unless the filters are cut artificially. On the other hand, our simple pinwheel framelets [5] and pinwheel framelets [7] are produced by finite length filters (or compactly supported functions). The Fig. 1 is the maximal overlap version of our pinwheel framelets of degree 7 at the second level.

3 Simulations of Visual Illusions

Based on the maximal overlap version of our pinwheel framelet we constructed a nonlinear model of visual information processing in the striate cortex. This model produces computer simulations of several lightness illusions in a unified way: This method covers the Hermann grid illusion, Mach band illusion, the Chevreul illusion etc. Here I show as an example simulations of the Hermann grid illusion and some variations.

The Hermann grid is shown in Fig. 2. In this image, small gray spots are perceived at the intersections of white bands, however there are no such spots. (Note that the illusion disappear in foveal vision.) Our simulation of the illusion is Fig. 3.

Fig. 2 Hermann grid

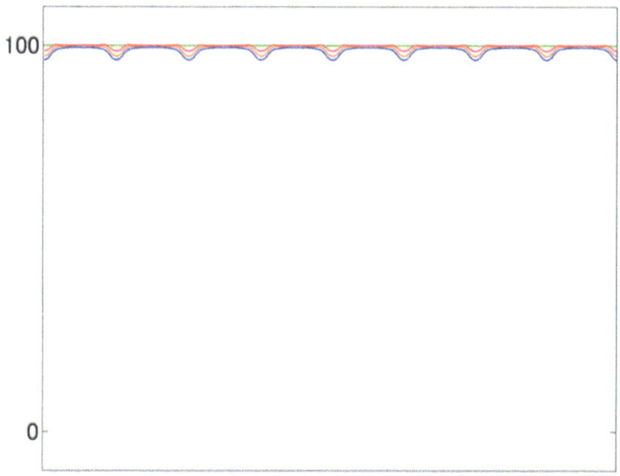

Fig. 3 Simulation of the Hermann grid illusion by using our nonlinear model with the pinwheel framelet of degree 5

So far it was thought that the Hermann grid illusion is caused by lateral inhibition in the retina ([8]). However as reported in Spillmann [11] and Wolfe [12], some evidences of a postretinal contribution to the illusion were found: the illusion is weakened when the grid is presented diagonally, and the strength of the illusion is effected by the number of intersections (see Fig. 4). Our mathematical model can simulate these phenomena. For example I show the simulation of the former phenomenon in Fig. 5.

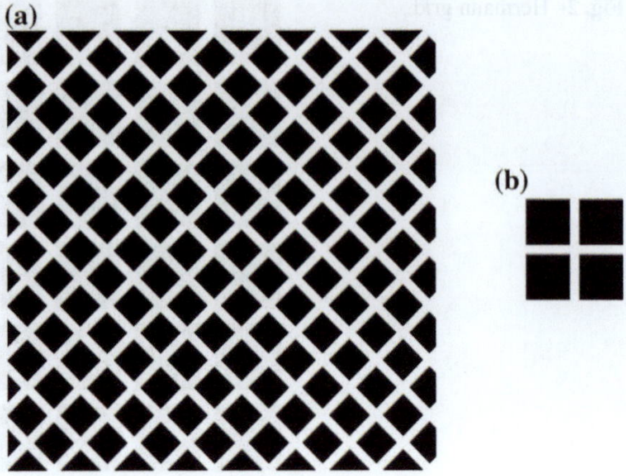

Fig. 4 **a** Diagonally presented Hermann grid, **b** Hermann grid with one intersection of white roads

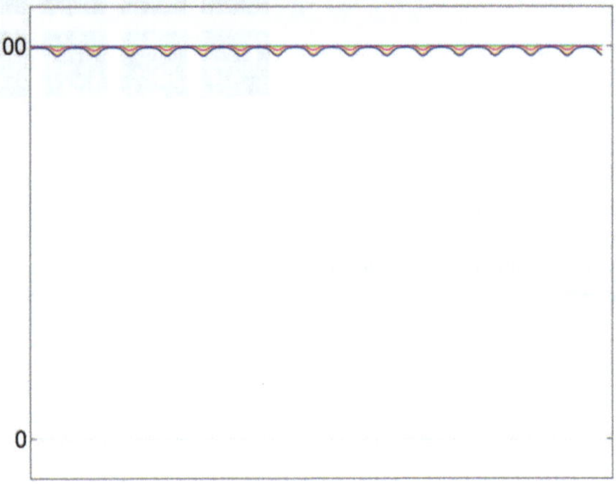

Fig. 5 Simulation of the diagonally presented Hermann grid illusion by using our nonlinear model with the pinwheel framelet of degree 5

4 Applications to Image Processing

Our nonlinear models of visual information processing in the brain have many applications not only to visual illusions, but also to image processing. A nonlinear model which has given us simulations of several lightness illusions and some illusions related to color perception provides a new method of sharpening of natural images. Our method can sharpene the parts a person generally wants to see without losing the overall image. For an example of our image processing, see [4]. This technique is patent pending (H. Arai and S. Arai, JST).

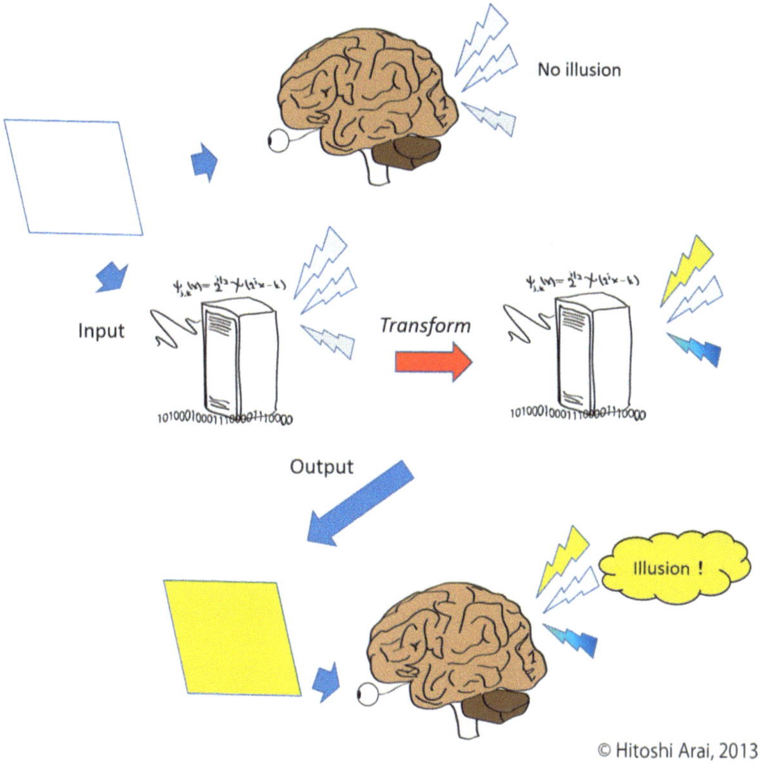

Fig. 6 Idea of transforming to fuyuu illusions

5 Creation of Visual Illusions

By one of our mathematical models of visual information processing, we can transform usual images or designs to certain illusory images: still images and designs seem to be moving, although they are at rest in the images (patent, H. Arai and S. Arai, JST). We call illusions produced by our method "fuyuu illusions". For examples of our fuyuu illusions, see [4] or the following site:

> Arai, H and Arai, S.: The Birth of "New Age Op Art",—new op art born from mathematical science of optical illusions (2011).
> http://www4.ocn.ne.jp/~arai/Exhibition/VisualIllusions.html

Why can our mathematical model create such illusions? A main idea is indicated in Fig. 6.

6 Other Results

We have other results related to mathematical models of visual information and applications to image processing. For example, mathematical algorithm of creating letter-row tilt illusions, noise reduction, edge detection, etc (patents, Arai and Arai, JST).

Acknowledgments The author thanks to Professor Anjyo for inviting me to Symposium MEIS2013.

References

1. Arai H (2010) Wavelets. Kyoritsu Publ, Tokyo (in Japanese)
2. Arai H (2005) A nonlinear model of visual information processing based on discrete maximal overlap wavelets. Interdiscip Inf Sci 11:177–219
3. Arai H (2007) Achromatic and chromatic visual information processing and discrete wavelets. In: Frontiers of Computational Sciences (pp 83–89, invited paper): Springer
4. Arai H (2013) Visual illusions and sciences. Minerva Shobo, Japan (in Japanese)
5. Arai H, Arai S (2009) 2D tight framelets with orientation selectivity suggested by vision science. JSIAM Lett 1:9–12 (Invited Paper)
6. Arai H, Arai S (2010) Framelet analysis of some geometrical illusions. Jpn J Ind Appl Math 27:23–46
7. Arai H, Arai S A patent specification (JP Patent No. 5038547; Intern. Pub. No. WO2012067254), a paper is in preparation
8. Baumgartner G (1960) Indirekte Größenbestimmung der rezeptiven Felder der Retina beim Menschen mittels der Hermannschen Gittertäuschung. Pflüger's Archiv für die gesamte Physiologie des Menschen und der Tiere 272:21–22
9. Daubechies I, Han B, Ron A, Shen Z (2003) Framelets: MRA-based constructions of wavelet frames. Appl Comput Harm Anal 14:1–46
10. Escalante-Ramírez E, Silván-Cárdenas J (2005) A multiresolution directional-oriented image transform based on gaussian derivatives. Signal Process Image Commun 20:801–812
11. Spillmann L (1994) The herman grid illusion: a tool for studying human perceptive field organization. Perception 23:691–708
12. Wolfe JM (1984) Global factors in the hermann grid illusion. Perception 13:33–40
13. Young RA (1991) Oh say, can you see? The physiology of vision. SPIE 1453:92–123

Decomposition and Clustering for the Visualization of Dynamical Systems

Zin Arai

Abstract We discuss recently developed methods for the visualization of dynamical systems which are based on the decomposition of the phase space of the system. The information of the system is represented by a directed graph and then the graph will be decomposed into smaller subsets which eventually defines a partition of the phase space. Depending on the purpose of visualization and the nature of the system, two different decompositions are introduced. The first decomposition algorithm is called Conley-Morse decomposition, which decompose the system according to the gradient-like structure of the system. On the other hand, the latter algorithm, an application of the peer pressure clustering algorithm for directed graphs, decompose each recurrent components of the system into further smaller non-invariant subsets according to the similarity of the dynamical behavior.

Keywords Dynamical systems · Strongly connected components · Conley-Morse decomposition · Graph clustering · Peer pressure clustering

1 Introduction

An effective visualization of the global behavior of a dynamical system or a fluid simulation inevitably involves a sort of the partition, or the decomposition of the phase space of the system. This is because, in a generic dynamical system there exist uncountably many orbits having "similar" dynamical behaviors. If we plot too many of them then typically we end up with a picture carrying no information. For example, Fig. 1 illustrates 300 (left) and 3,000 (right) different trajectories of *the standard map*, the most important example of Hamiltonian dynamics, of length 100.

Z. Arai (✉)
Department of Mathematics, Hokkaido University/JST CREST,
Kita 10 Nishi 8, Kita-ku, Sapporo 060-0810, Japan
e-mail: zin@math.sci.hokudai.ac.jp

K. Anjyo (ed.), *Mathematical Progress in Expressive Image Synthesis I*,
Mathematics for Industry 4, DOI: 10.1007/978-4-431-55007-5_3,

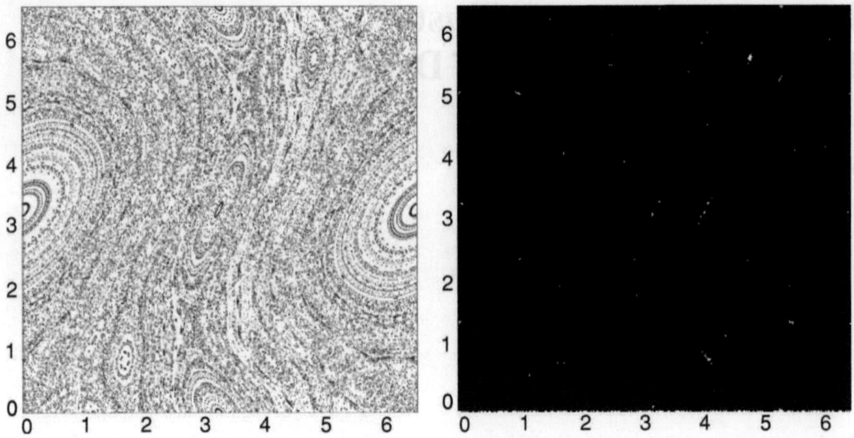

Fig. 1 Plots of 300 (*left*) and 3,000 (*right*) orbits of $(x, y) \mapsto (x + k \sin y, x + y + k \sin y)$, the standard map with the parameter $k = 0.971635$. The map is

It easy to imagine that this problem will be more serious in higher dimensional spaces where we can not use our geometric and physical intuition on the space.

Thus, a natural way to visualize the global behavior of the systems is to classify the points in the phase space into a relatively small number of clusters, so that each cluster corresponds to a particular dynamical behavior.

In following sections, we will explain two recently proposed ideas for obtaining such a clustering of dynamical systems. In the next section, we will briefly review the Conley-Morse decomposition, a decomposition of the phase space according to the gradient-like structure of the system. Unfortunately this method does not work fine for conservative dynamics such as Hamiltonian dynamics and hence we will discuss another algorithm based on a graph clustering algorithm in the last section. This is a work in progress in the JST CREST project [9].

2 Conley-Morse Decomposition

In this section, we discuss the method of the Conley-Morse decomposition [1, 2]. The key idea here is to find small subsets in the phase space which are invariant under the dynamics, and then decompose the phase space into these subsets and connecting orbits among them. These invariant subsets will be called Morse sets.

Historically, this idea was first applied to the gradient flows satisfying a certain non-degeneracy condition by M. Morse. Here by a gradient flow we mean a flow on a manifold M that is defined by the gradient vector field grad f for a smooth function $f : M \to \mathbb{R}$. Note that in a gradient flow, there will be no periodic orbit (or, closed orbit) nor chaotic orbit and thus Morse sets are just equilibrium points of the flow.

Fig. 2 Collapsing strongly connected components of G

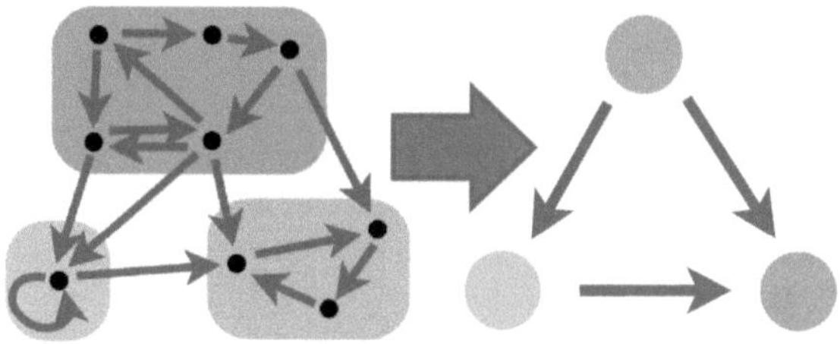

Note that equilibrium points of the flow generated by grad f are exactly the critical points of f.

Then C. Conley generalized the theory to arbitrary continuous dynamical systems yielding the celebrated Fundamental theorem of dynamical systems, which says that a dynamical system can always be decomposed into (possibly chaotic) Morse sets and non-chaotic connecting orbit among them. In this generalized case, since the dynamics is not assumed to be a gradient flow anymore, a Morse sets can be an equilibrium point, a periodic orbit, or a chaotic invariant set such as the Lorenz attractor. The fundamental theorem does not tell the detail of the dynamics inside Morse sets (except the fact they should satisfy a sort of recurrence condition called *chain recurrence*), but the point is that we can clearly separate chaotic and non-chaotic region of the dynamics.

When the dynamics enjoys the structural stability (essentially equivalent to the property called uniform hyperbolicity), there exists finitely many Morse sets and the dynamics on them can be described in a symbolic manner (Markov partition). Unfortunately, there may be infinitely many Morse sets in general and in such a case, the gradient structure outside the Morse sets will not be robust under small perturbations. However, since we are mainly interested in the application to practical problems in which noise and errors are inevitably involved, we can restrict ourselves to finitely many larger Morse sets. In practice, this can be achieved by fixing the grid size for our computation and ignoring Morse sets smaller than the grid size.

Given a dynamical system and a grid decomposition of the phase space, the first step of the algorithm is to define a graph G whose edges imitate the dynamics. Usually, we simply subdivide the phase space into small cubes using a uniform grid, and then encode the dynamical behavior by looking how the image of each cube intersects other cubes (see Arai et al. [2, 4], for example).

Then we can expect that a Morse set corresponds to a *strongly connected component* in G. Given a directed graph G, a subset of vertices of G is called strongly connected if for any v, w in the set, there exist directed paths from v to w and w to v. It is well knows that there exist linear-time algorithms that can decompose a directed graph G into strongly connected components.

By collapsing each strongly connected component of G to a single node, we can obtain a much smaller graph $\mathscr{G}$ representing the structure of the Conley-Morse decomposition (Fig. 2). Note that G could be very huge depending on the dimension of the phase space and the size of the grid we are using, however, the graph $\mathscr{G}$

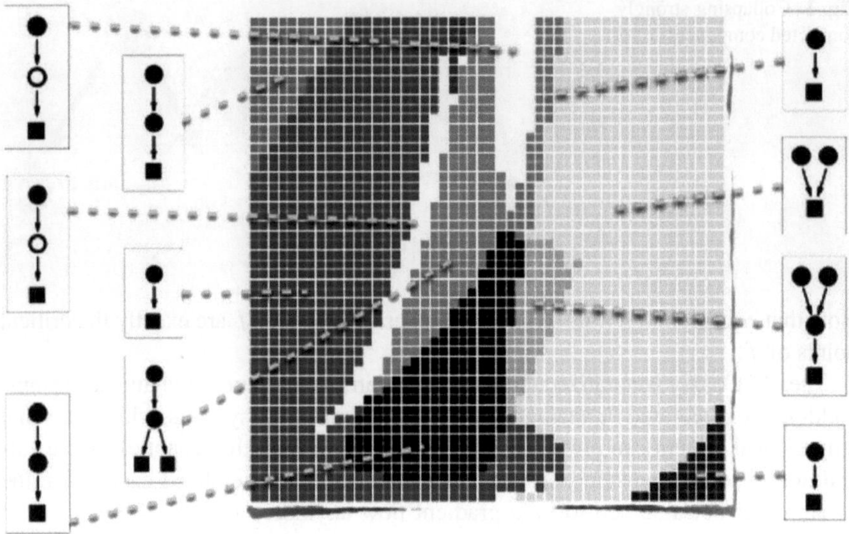

Fig. 3 The "bifurcation" diagram of the Leslie population model

obtained after collapsing would be much smaller than G. A node in the collapsed graph corresponds to a Morse set of the system.

For each Morse set, we then compute *the Conley index*, which is an algebraic topological invariant carrying the information of the dynamics in a neighborhood of the Morse set [3, 6]. Finally, for each vertex of $\mathcal{G}$ we associate the Conley index of the corresponding Morse set. The obtained data structure is called *the Conley-Morse graph* and the corresponding decomposition of the system is called *the Conley-Morse decomposition*.

Figure 3 illustrates an example of the application of Conley-Morse decomposition to the Leslie population model, a map defined by

$$f(x_1, x_2; \theta_1, \theta_2) = ((\theta_1 x_1 + \theta_2 x_2) \cdot e^{-0.1(x_1+x_2)}, 0.7x_1).$$

The map defines a dynamical system $f : \mathbb{R}^2 \to \mathbb{R}^2$ with two parameters θ_1, θ_2 that represent the birth rate of the first and second generation, respectively. The figure shows a decomposition of the parameter (θ_1, θ_2)-plane according to the obtained Conley-Morse graph structure; adjacent boxes in the parameter space with equivalent Conley-Morse graphs are plotted in the same color. A square, a filled circle, a hollow circle in directed graphs indicates an attractor, a Morse set with non-trivial Conley index, and a Morse set with trivial Conley index, respectively. If the Conley index of a Morse set is non-trivial, then we can prove the Morse set is non-empty, that is, there really exist some orbit completely contained in the Morse set. On the other hand, if the Conley index is trivial, the Morse set might be empty; typically, this

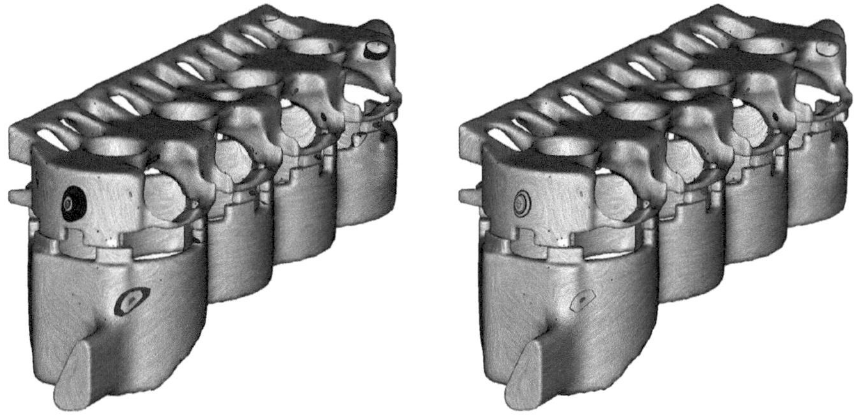

Fig. 4 Morse sets for a flow on the surface of a cooling jacket

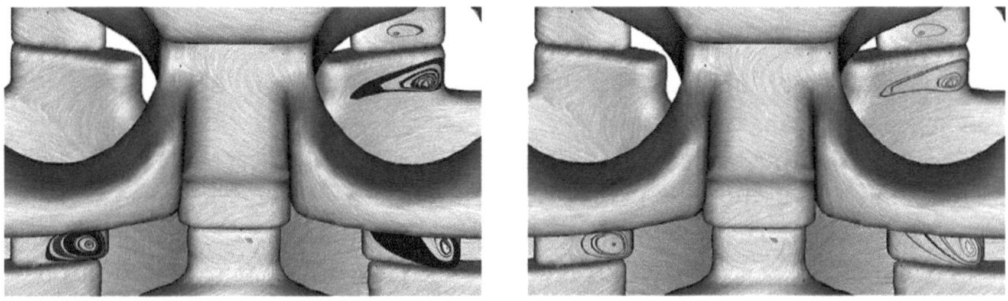

Fig. 5 A close up view of Morse sets

happens when our gird size is too coarse to distinguish a very slow dynamics from a true invariant set.

Although the algorithm works fine for the Leslie model and some other lower dimensional problems, its computational cost is still expensive to be applied to more practical problems. To overcome this computational difficulty, Szymczak et al. developed the method of the piecewise constant approximation of a vector field [7, 8] using the theory of differential inclusions. The trajectories in a piecewise constant vector filed can be determined by simple geometric rules and hence we can avoid computationally expensive numerical integrations.

Figures 4 and 5 show the result of the application of Szymczak's idea to the vector field on the surface of a cooling jacket which is induced by extrapolating data from 3D fluid simulation in the jacket (figures are provided by the courtesy of A. Szymczak). Note that the original 3D flow does not have any attractors nor repellers since it is incompressible, but the flow on the surface has a gradient-like structure illustrated in the figures, which represents the dynamics vertical to the surface.

3 Graph Clustering Algorithms

The idea of Conley-Morse decomposition that we have seen in the previous section works generally well for dissipative dynamics, in which we can expect the existence of attractors and repellers. On the other hand, in conservative systems, for example in Hamiltonian dynamics, there is no attractor and repeller and thus typically the Conley-Morse decomposition becomes a trivial one.

To obtain an effective visualization of conservative systems, therefore, another criterion for the phase space partition is required. Note that if the system $T : X \to X$ is ergodic, we can not use a function $f : X \to \mathbb{R}$ to define such a partition of X because for almost all initial point $x \in X$, the time average

$$\lim_{n\to\infty} \frac{1}{n} \sum_{i=0}^{i=n} f(T^i(x))$$

takes the same value. This suggests that our criterion should be based on the finite-time behavior of trajectories, rather than the asymptotic behavior as we have done in the Conley-Morse decomposition.

Here we propose a practical algorithm based on a graph clustering algorithm called Peer Pressure Clustering (PPC) [5]. A graph clustering algorithms classifies the nodes of a graph with respect to some similarity between nodes, where the similarity is defined suitably depending on the purpose of the clustering. These algorithms have been applied to the study of social networks, machine learning, data mining, pattern recognition, image analysis, and bioinformatics.

At the initial condition of PPC algorithm, each node of the graph compose a cluster to which only this node belongs. Then PPC iteratively refines the clustering so that the connectivity inside a cluster will be higher than the connectivity between different clusters. More precisely, in every iteration step, each node of the graph decides to which cluster it should belong by feeling the pressure from its friends (adjacent nodes); if there are many adjacent nodes that belong to a particular cluster, then the node choose to belong the same cluster. Unfortunately, the convergence of the algorithm is not guaranteed for a general graph; it may end up with a periodic emergence of different clustering. However, in most of our cases the algorithm stops after a small number of iterations and runs much faster than other clustering algorithms such as Markov clustering.

In what follows, we apply PPC algorithm to the graph G obtained from a given dynamics. The resulting clustering of G corresponds to a partition of the phase space of the system. Since G imitates the behavior of the original dynamics, we can expect that this partition of the phase space enjoys a property similar to the clustering of G; the mixing inside a cluster should be stronger than mixing between different clusters.

Figure 6 shows the result of PPC for a graph obtained from the standard map. We note that quasi-periodic motions and chaotic motions are clearly separated by the algorithm. We note that the directed graph obtained from the system is itself strongly

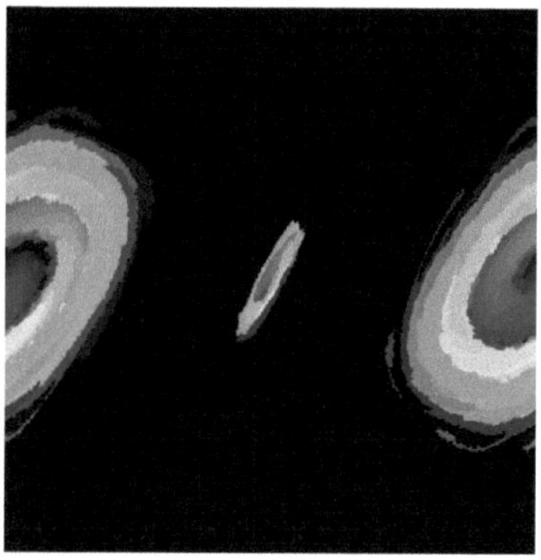

Fig. 6 A clustering of the phase space of the standard map obtained by PPC algorithm

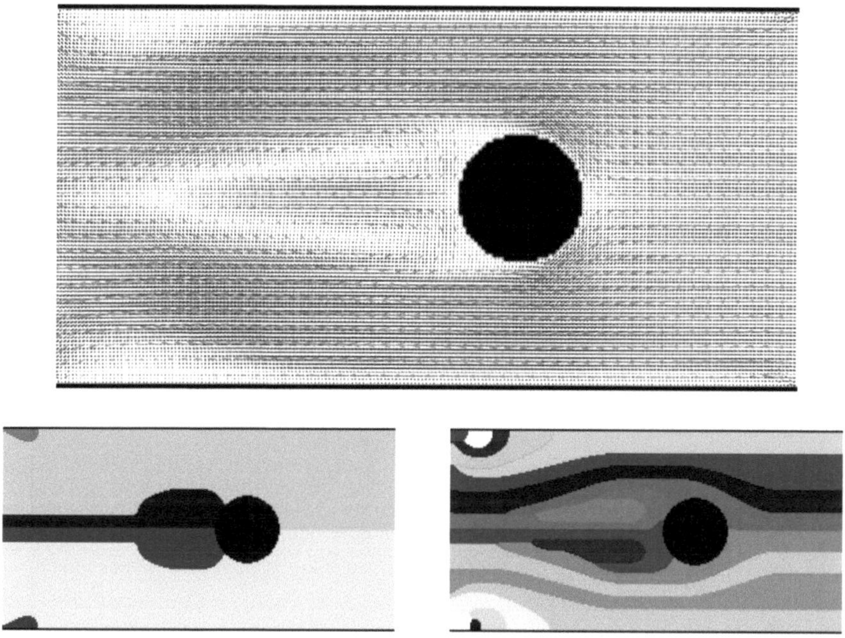

Fig. 7 The flow (*top*) and the results of coarser (*bottom left*) and finer (*bottom right*) PPC clustering

connected, and hence the Conley-Morse decomposition algorithm gives the trivial decomposition, that is, the only non-empty Morse set is the entire phase space.

Figure 7 shows a 2D numerical fluid simulation using the Lattice-Boltzmann method and the result of PPC applied to it. Two PPC results are shown for different values of a parameter in the algorithm that control the granularity of the clustering.

References

1. Arai Z, Kalies W, Kokubu H, Mischaikow K, Oka H, Pilarczyk P (2009) A database schema for the analysis of global dynamics of multiparameter systems. SIAM J Appl Dyn Syst 8(3):757–789
2. Arai Z, Kokubu H, Pilarczyk P (2009) Recent development in rigorous computational methods in dynamical systems. Japan J Ind Appl Math 26(2–3):393–417
3. Conley C (1978) Isolated invariant sets and the Morse index, CBMS Regional Conference Series in Mathematics, vol 38. American Mathematical Society, Providence, RI
4. Dellnitz M, Junge O (2002) Set oriented numerical methods for dynamical systems? Handbook of dynamical systems, vol 2. North-Holland, pp. 221–264
5. Kepner J, Gilbert J (eds) (2011) Graph algorithms in the language of linear algebra. Society for Industrial and Applied Mathematics, Philadelphia, PA
6. Mischaikow K, Mrozek M (2002) Conley index. Handbook of dynamical systems, vol 2. North-Holland, pp 393–460
7. Szymczak A, Zhang E (2012) Robust Morse decompositions of piecewise constant vector fields. IEEE Trans Vis Comput Graph 18(6):938–951
8. Szymczak A (2013) Morse connection graphs for piecewise constant vector fields on surfaces. Comput Aided Geom Des 30(6):529–541
9. The CREST Project Toward a paradigm shift created by mathematics of vortex-boundary interactions (2014). http://www.math.sci.hokudai.ac.jp/crest/

Probable and Improbable Faces

J. P. Lewis, Zhenyao Mo, Ken Anjyo and Taehyun Rhee

Abstract The multivariate normal is widely used as the expected distribution of face shape. It has been used for face detection and tracking in computer vision, as a prior for facial animation editing in computer graphics, and as a model in psychological theory. In this contribution we consider the character of the multivariate normal in high dimensions, and show that these applications are not justified. While we provide limited evidence that facial proportions are not Gaussian, this is tangential to our conclusion: even if faces are truly "Gaussian", maximum a posteriori and other applications and conclusions that assume that typical faces lie near the mean are not valid.

Keywords Principal component analysis (PCA) · Multivariate Gaussian (distribution) · Blendshapes · Mahalanobis distance · Curse of dimensionality · Active appearance model (AAM) · Maximum a posteriori (MAP)

J. P. Lewis (✉)
School of Engineering and Computer Science,
Victoria University of Wellington/JST CREST, PO Box 600, Wellington 6140,
New Zealand
e-mail: jplewis@ecs.vuw.ac.nz

Z. Mo
Google Inc., 1600 Amphitheatre Parkway, Mountain View, CA 94043, USA
e-mail: zhenyao@gmail.com

K. Anjyo
OLM Digital, Inc./JST CREST, Dens Kono Bldg. 302, 1-8-8,
Wakabayashi, Setagaya-ku, Tokyo 154-0023, Japan
e-mail: anjyo@olm.co.jp

T. Rhee
School of Engineering and Computer Science, Victoria University of Wellington,
PO Box 600, Wellington 6140, New Zealand
e-mail: taehyun.rhee@ecs.vuw.ac.nz

K. Anjyo (ed.), *Mathematical Progress in Expressive Image Synthesis I*,
Mathematics for Industry 4, DOI: 10.1007/978-4-431-55007-5_4,

1 Introduction

In computer vision and graphics research, facial expression and identity are commonly modeled as a high-dimensional vector space, often with a multidimensional Gaussian density. This choice of representation has associated algorithmic approaches such as linear interpolation and maximum a posteriori (MAP) solution of inverse problems.

In this chapter we argue several things: (1) the linear and Gaussian assumptions are not strictly correct. (2) Existing research that starts from these assumptions has implicitly assumed a low dimensional setting. In high dimensions, common algorithmic approaches such as MAP may not be justified. (3) The problems resulting from these assumptions are not just hypothetical, but are visible in a practical computation, specifically interpolation of faces. Most importantly, we show that consideration of these factors can result in an algorithm with visibly improved results.

2 Linear Models

The faces of realistic computer characters in movies are most often generated using the "blendshape" representation [1, 4, 5, 13]. This is a linear representation of the form $\mathbf{f} = \mathbf{B}\mathbf{w}$, where $\mathbf{B}$ is a linear but non-orthogonal basis having semantic meaning. In computer vision, approaches such as active appearance models (AAM) [3] and morphable models [2] use an orthogonal basis generated by principal component analysis (PCA), and assume the multidimensional Gaussian prior. Bilinear (tensor) face models have also been proposed [15]. Psychological research has also employed such linear models with a multivariate Gaussian prior [14].

PCA assumes that the data is jointly Gaussian, in that the PCA basis vectors are the eigenvectors of a covariance matrix that does not capture any non-Gaussian statistics. The Gaussian assumption leads to a frequently employed prior or regularizer of the form $\mathbf{c}^T \boldsymbol{\Lambda}^{-1} \mathbf{c}$ where $\mathbf{c}$ is the vector of PCA coefficients and $\boldsymbol{\Lambda}$ is the diagonal matrix of eigenvalues (variances). The Gaussian assumption also naturally leads to the MAP approach to regularising inverse problems. This approach selects model parameters M as the mode of the posterior $P(D|M)P(M)$ given data D. With a Gaussian model the posterior also has a Gaussian form.

The appropriate number of dimensions for a linear facial model of expression or identity has been variously estimated to be in the range 40–100 [7, 8, 11]. High quality blendshape facial models used in movie visual effects often have on the order of 100 dimensions [4].

In Fig. 1 we show that the common multidimensional Gaussian assumption is not strictly accurate. This figure shows a kernel density plot of several simple measurements of facial proportions measured from 359 selected photographs from the facial database [12]. It is also somewhat obvious that a linear model is not entirely appropriate for facial *expression*. For example, the motion of the jaw has a clear rotational

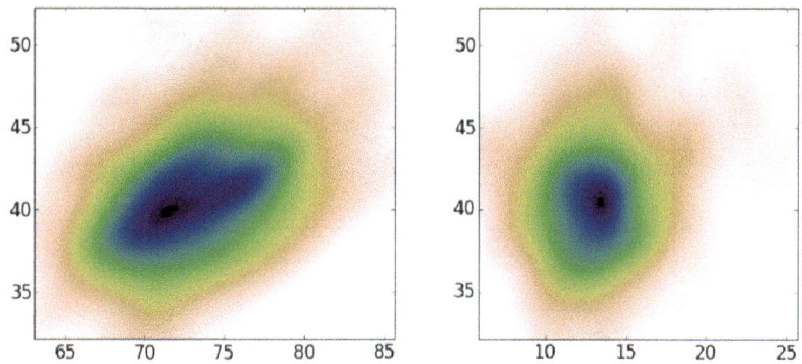

Fig. 1 Face proportions are not strictly Gaussian. Kernel density plots of (*left*) the distance between the eyes versus the width of the mouth, (*right*) the width of the mouth versus the height of the mouth, measured from a database of 359 faces

component. On the other hand, the widespread use of the blendshape representation in movies (albeit sometimes with nonlinear correction terms [13]) is an argument that linear models suffice even if they are not strictly accurate. It is less clear whether a vector space model of facial *identity* is appropriate, or if a (nonlinear) manifold assumption would be more accurate. While these comments call into question the linear and Gaussian assumptions, existing research does not indicate whether these objections are important in practical computations.

3 High-Dimensional Phenomena

High dimensional data is generally subject to a collection of nonintuitive phenomena collectively known as the "curse of dimensionality" [16]. Examples of such phenomena are that (a) in high dimensions, "all data is far away" with high probability (Fig. 2), (b) randomly chosen vectors are nearly orthogonal (Fig. 3), and (c) the probability mass of the data is overwhelmingly located near the surface of the hypervolume, with the interior of the volume essentially empty (Figs. 4 and 5).

Current face computation approaches generally overlook these phenomena. A notable exception is [10], who described the following apparent paradox: the squared Mahalanobis distance $\mathbf{c}^T \mathbf{\Lambda}^{-1} \mathbf{c}$ follows a χ^2 distribution with n degrees of freedom, since it is the sum of i.i.d. squared Gaussian variables of variance $\frac{c_i^2}{\lambda_i}$. The expectation of this distribution for d dimensions is d, thus we expect the length of the standardized squared coefficient vector of a typical face to be d. However under the multidimensional Gaussian model, the face at the origin (the mean face) is the most probable, and the length of its squared coefficient vector is zero.

Patel and Smith [10] also state a hypothesis that faces should lie on the shell of a hyperellipsoid dictated by the squared coefficient length. The resolution to the apparent paradox is simply that it is the difference between the variance and mean.

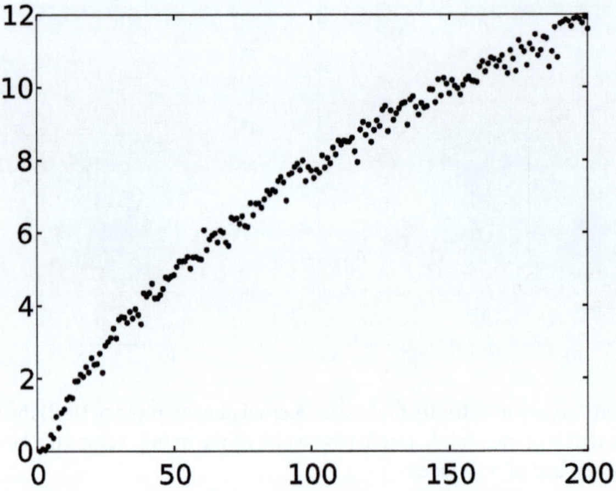

Fig. 2 The closest distance to the mean among 1,000 unit-variance multidimensional Gaussian random variables (*vertical axis*) as a function of the dimension (*horizontal axis*). In 100 dimensions every point in this simulation is more than six standard deviations from the mean

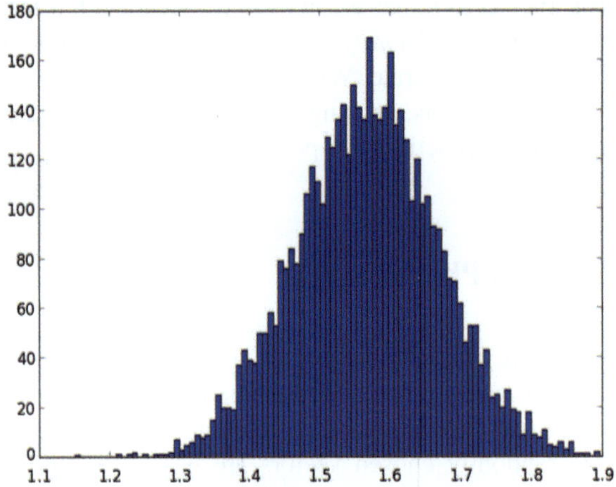

Fig. 3 In high dimensions, most data are nearly orthogonal. Histogram of the angles between all pairs of 100 randomly chosen isotropic Gaussian random variables in 100 dimensions. The angles cluster around $\pi/2$

A zero-mean random variable can (and typically does!) have a nonzero variance. Randomly sampling from a multidimensional Gaussian will generate a sequence of samples that have *both* the expected mean and variance of course.

Fig. 4 *Bottom* the one dimensional Gaussian distribution, with the area between one and two deviations indicated in *red*. This interval is equal to that of the unit radius. *Top* In two dimensions, the area between one and two standard deviations (*light blue*) is relatively larger than the area of the unit standard deviation disc (*light orange*). Figure is best viewed in the electronic version of this document

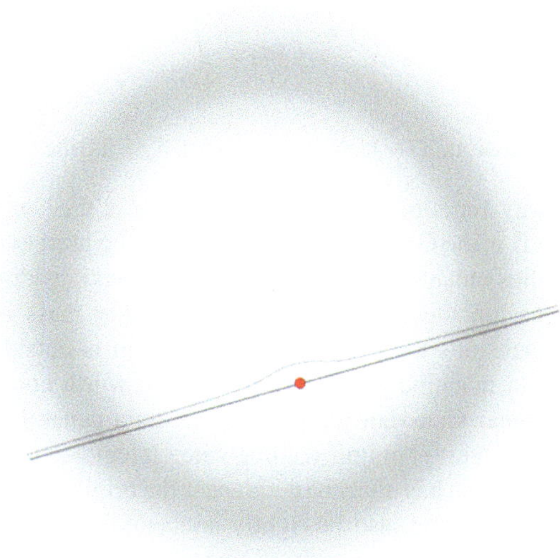

Fig. 5 The point along the constraint (*dark line*) that has the highest probability is the *red point*. In high dimensions however, the interior of the Gaussian is empty and the probability mass is concentrated toward the outside

4 Discussion

Next we will verify the statement that high dimensional data is concentrated overwhelmingly near the surface of the hypervolume. In the case of a uniformly distributed random variable in a hypercube, this is easy to see. Consider a unit hypercube in d

dimensions, that encloses a smaller hypercube of side $1-\varepsilon$. As $d \to \infty$, the volume of the enclosed hypercube is $(1-\varepsilon)^d \to 0$.

The fact that the multivariate Gaussian is a heavy tailed distribution in high dimensions is less obvious. For example, [14] states, "even for a face space of high dimensionality, the assumption of a multivariate normal distribution means that... There will be many typical faces that will be located relatively close to the center". However this phenomenon is at least intuitively suggested by comparing the one- and two-dimensional Gaussian distributions (Fig. 4). In one dimension, the "volume" of the interval between one and two standard deviations is equal to the radius of the unit interval. In two dimensions the area of the annulus between one and two standard deviations is relatively larger than the area of the unit disc. In higher dimensions the trend continues, with the available volume overwhelmingly concentrated near the outside.

Discussion of the multivariate Gaussian is simplified by a "whitening" transformation $c_i \to c_i/\sqrt{\lambda_i}$ from the original hyperellipsoidal density to an isotropic density. We can also consider a unit-variance density without loss of generality. In this case the probability that a point is within a hypersphere of radius r is proportional to

$$\int_0^r S_{d-1}(r)G(r) = \frac{2\pi^{d/2}}{\Gamma(d/2)} \int_0^r r^{d-1}G(r)dr$$

where d is the dimension, $G(r) = \frac{1}{\sqrt{(2\pi)^d}} \exp^{-r^2/2}$ is the isotropic unit variance Gaussian density function, $S_{d-1}(r) = \frac{2\pi^{d/2}r^{d-1}}{\Gamma(d/2)}$ is the "surface area" of the d-hypersphere, and Γ is the Gamma function. This can be used to plot the tail probability that a point lies outside the unit hypersphere in various dimensions (Fig. 6). While in one dimension the majority of the probability mass is within the unit interval, in 100 dimensions the probability that a point is outside the unit hypersphere is 1. to within machine precision! It may be worth contrasting the mode of the high-dimensional Gaussian with the Dirac delta generalised function. The delta function has zero width but unit volume when integrated over. In contrast, the high-dimensional Gaussian has nonzero width near the origin, but negligible volume.

High dimensional data can also be tightly concentrated in a shell of relatively narrow thickness. In the case of the multi-dimensional Gaussian, the majority of its mass is concentrated within a shell centered at radius $\sqrt{d}$. Figure 7 plots the radially integrated unit variance Gaussian profile $S_{d-1}(r)G(r)$ relative to the distance $\sqrt{d}$ (i.e. with a change of variable $r \to r\sqrt{d}$). The data is concentrated increasingly around $\sqrt{d}$ (relative to the distance $\sqrt{d}$ itself) in high dimensions.

The observations collected above lead to the remarkable conclusion that algorithms such as MAP may be nonsensical in high dimensions! This conclusion is not widely known in the computer vision and graphics community, where MAP is commonly used for face computations with models having 10–100 dimensions.[1]

[1] In fact many results in statistics focus on the case where increasing amounts of data are available, i.e. $n/d \to \infty$ with n the number of data points. In our problem we may have n/d finite and small,

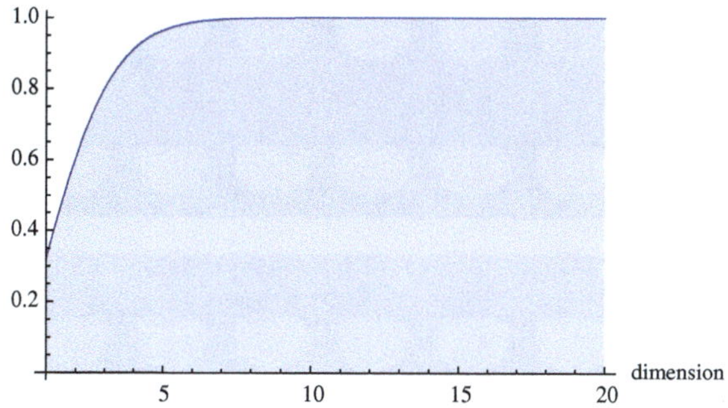

Fig. 6 Probability that a sample from a unit variance Gaussian is outside the unit hypersphere for various dimensions

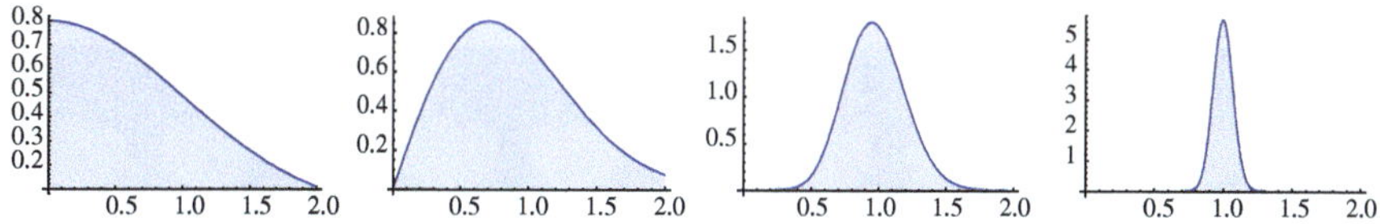

Fig. 7 The radially integrated Gaussian $N(0, \mathbf{I}_n)$ in various dimensions. Each subfigure shows the radially integrated Gaussian profile $S_{d-1}(r)G(r)$ (*vertical axis*) plotted in units of $\sqrt{d}$ (*horizontal axis*). *From left to right* 1, 2, 10, and 100 dimensions. In high dimensions the probability concentrates in a shell centered at radius $\sqrt{d}$

However, our conclusion is supported in [6], where Mackay states "probability density maxima often have very little associated probability mass even though the value of the probability density there may be immense, because they have so little associated volume... the locations of probability density maxima in many dimensions are generally misleading and irrelevant. Probability densities should only be maximized if there is good reason to believe that the location of the maximum conveys useful information about the whole distribution."

5 Example Computation: Interpolating in Face Space

Figure 8 contrasts two approaches to interpolating facial identity. The images are not photographs but are synthesized with an AAM [9]. The face on the far left is generated from a coefficient vector $\mathbf{c}_l$ sampled from a multivariate Gaussian with the appropriate variances (eigenvalues). The face on the far right is also randomly chosen, but its coefficient vector $\mathbf{c}_r$ is modified to constrain it to having a specified inner product $\langle \mathbf{c}_l, \mathbf{c}_r \rangle_{\Lambda^{-1}} = -0.8$ so as to place it on the opposite side of the coefficient volume. The inner product uses the inverse eigenvalue-weighted norm $\langle \mathbf{c}_l, \mathbf{c}_r \rangle_{\Lambda^{-1}} = \mathbf{c}_l^T \mathbf{\Lambda}^{-1} \mathbf{c}_r$. The dimensionality of the space (length of the coefficient vector) is 181.

The top rows in Fig. 8 shows linear interpolation through the Gaussian coefficient space. The midpoint of this interpolation passes closer to the center (mean) face than

as in the case of a face model with several hundred training examples, each with 100 degrees of freedom.

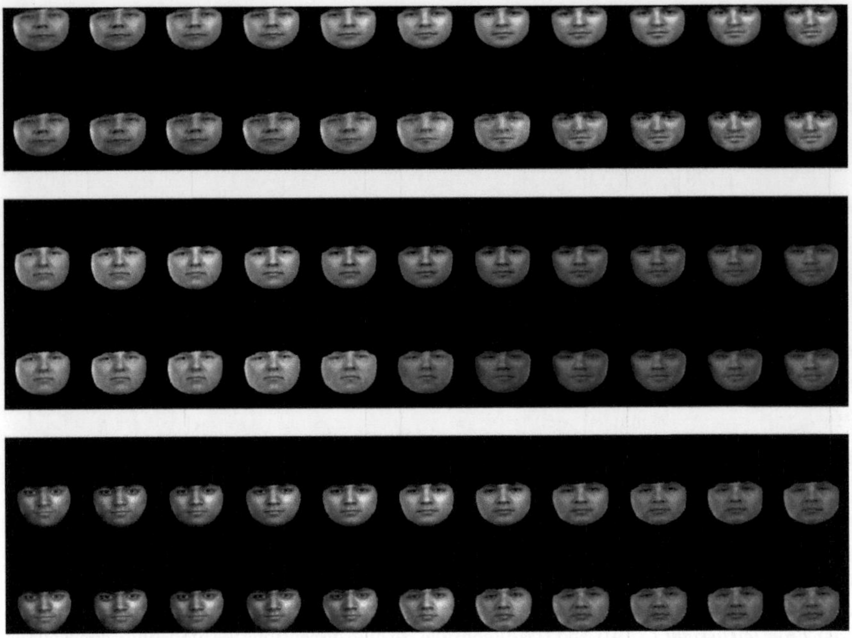

Fig. 8 Interpolating between a randomly chosen face (*left column*) and a second face (*right column*) nearly on the opposite side of the hyperellipse of coefficients. *Top row* of each image: linear interpolation of coefficients. The *middle* images lack distinctiveness. *Bottom row* of each image: interpolating "around the hyperellipse". Detail is preserved throughout the interpolation. Please enlarge to see details

either end. This results in a somewhat "ghostly" face that lacks detail. The linear interpolation also has the undesired result that (for example) interpolating from a person of age 40 to a person of age 45 might pass through an intermediate face of apparent age 25, if that is the mean age of the database underlying the AAM.

In the lower panels of Fig. 8 we interpolate "around" a hyperellipsoidal shell in the coefficient space rather than across the volume. Given initial and final coefficient vectors $\mathbf{c}_l$, $\mathbf{c}_r$, at each step a coefficient vector is generated that interpolates the norm of these vectors (although in fact the difference in norm is expected to be small due to phenomena mentioned above). This interpolation remains inside the high probability shell of the hyperGaussian and generates distinctive faces throughout the interpolation.

6 Conclusion

This chapter describes known high-dimensional phenomena that call into question common assumptions underlying much computer vision, graphics, and psychological research on face computation. In particular, we question approaches that assume that typical faces lie in the interior of a high-dimensional Gaussian density (Fig. 5). The issue is not due to a bi- or multimodal distribution (as with a combined distribution containing both women and men) but rather is a consequence of high dimensionality. These objections are not merely hypothetical, but are visible in a simple face computation. Our conclusion highlights the need to develop new algorithms that address the intrinsically high-dimensional nature of facial identity and expression.

Acknowledgments This research is partially supported by the Japan Science and Technology Agency, CREST project. JPL acknowledges a helpful discussion with Marcus Frean.

Appendix: Hyperellipsoidal Angle Calculation

The interpolations in Fig. 8 start with a randomly chosen coefficient vector $\mathbf{y}$ with $y_i \sim N(0, \sqrt{\lambda_i})$. This produces the first face. For the second face, we select a coefficient vector $\mathbf{x}$ that has a specified Mahalanobis inner product with that of the first face, $\mathbf{x}\boldsymbol{\Lambda}^{-1}\mathbf{y} = c$ with $c = -0.8$ for example. To find $\mathbf{x}$ we solve a sequence of problems

$$\mathbf{x} \leftarrow \arg\min_{\mathbf{x}} \quad (\mathbf{x}-\mathbf{r})^T\boldsymbol{\Lambda}^{-1}(\mathbf{x}-\mathbf{r}) \quad + \quad \lambda(\mathbf{x}^T\boldsymbol{\Lambda}^{-1}\mathbf{y}-c)$$

$$\mathbf{r} \leftarrow \frac{\mathbf{x}}{\mathbf{x}^T\boldsymbol{\Lambda}^{-1}\mathbf{x}}$$

with $\mathbf{r}$ initialized to a random vector, in other words, find the vector that is closest to $\mathbf{r}$ and has the desired Mahalanobis angle with $\mathbf{y}$.

References

1. Anjyo K, Todo H, Lewis J (2012) A practical approach to direct manipulation blendshapes. J Graph Tools 16(3):160–176
2. Blanz T, Vetter T (1999) A morphable model for the synthesis of 3d faces. In: Proceedings of ACM SIGGRAPH, pp 187–194
3. Cootes TF, Edwards GJ, Taylor CJ (1998) Active appearance models. In: Burkhardt H, Neumann B (eds) ECCV'98: computer vision. Proceedings of the 5th European conference on computer vision, Volume II. Lecture notes in computer science vol 1407, Springer, Berlin
4. Lewis J, Anjyo K (2010) Direct manipulation blendshapes. Comput Graph Appl (special issue: Digital Human Faces) 30(4):42–50
5. Li H, Yu J, Ye Y, Bregler C (2013) Realtime facial animation with on-the-fly correctives. ACM Trans Graph 42:1–10
6. MacKay DJ (1996) Hyperparameters: Optimize, or integrate out? In: Heidbreder G (ed) Maximum entropy and Bayesian methods. Springer, New York, pp 43–59
7. Matthews I, Xiao J, Baker S (2006) On the dimensionality of deformable face models. CMU-RI-TR-06-12

8. Meytlis M, Sirovich L (2007) On the dimensionality of face space. IEEE Trans Pattern Anal Mach Intell 29(7):1262–1267
9. Mo Z, Lewis J, Neumann U (2004) Face inpainting with local linear representations. In: BMVC, BMVA, pp 347–356
10. Patel A, Smith W (2009) 3D morphable face models revisited. In: Computer vision and pattern recognition (CVPR), IEEE Computer Society, Los Alamitos, CA, USA, pp 1327–1334
11. Penev PS, Sirovich L (2000) The global dimensionality of face space. In: Proceedings of 4th international conference automatic face and gesture recognition, pp 264–270
12. Phillips PJ, Wechsler H, Huang J, Rauss P (1998) The feret database and evaluation procedure for face recognition algorithms. Image Vis Comput J 16(5):295–306
13. Seo J, Irving G, Lewis JP, Noh J (2011) Compression and direct manipulation of complex blendshape models. ACM Trans Graph 30(6):164:1–164:10
14. Valentine T (2012) Face-space models of face recognition. In: Wenger M, Townsend J (eds) Computational, geometric, and process perspectives on facial cognition: contexts and challenges, Scientific Psychology Series. Taylor & Francis, Oxford
15. Vlasic D, Brand M, Pfister H, Popovic J (2005) Face transfer with multilinear models. ACM Trans Graph 24(3):426–433
16. Wang J (2011) Geometric structure of high-dimensional data and dimensionality reduction. Springer-Verlag, Berlin, Heidelberg

Part II
Sound and Scene Rendering

Progress in Digital Sound Synthesis for Physically Based Animation (Invited Talk)

Doug L. James

Keywords Sound synthesis · Sound rendering · Vibration analysis · Thin-shell solids · Brittle structure

Natural sounds are all around us. The sound of my son struggling to get out of his wet rain coat, or rain boots squeaking across the floor. The sound of lemonade pouring into an ice-filled glass, or the ocean crashing at your feet. The sound of an agitated shopping cart plunging down a flight of stairs, or the familiar roar of a camp fire. Reality produces these sounds "for free," but how can we best synthesize them in future computer-simulated realities?

Decades of advances in computer graphics and physics-based simulation have made it possible to convincingly animate a wide range of phenomena, such as contacting rigid and deformable bodies, fracturing solids, splashing water, and roaring fire. Such simulations will inevitably run in real time one day, paving the way for interactive virtual environments. Unfortunately, the realities simulated by current algorithms are essentially "silent movies," with sound added as an afterthought. Instead, sound recordings have been edited manually for pre-produced animations, or triggered automatically in interactive settings. The former is labor intensive and inflexible; and the latter may produce awkward, repetitive and/or implausible results. Prior synthesis techniques lack the sophistication to sonify increasingly sophisticated physics-based animations. This situation is a serious obstacle to our ultimate goal: to build wonderous, real-time or off-line, multi-sensory experiences on future hardware platforms where graphics, motion, and sound are synchronized and highly engaging.

In this talk, I will describe progress toward these synthesis goals by our group at Cornell, and mention some of the many remaining challenges. I will discuss progress on modeling sound for visually important phenomena such as rigid bodies [4, 5],

D. L. James (✉)
Cornell University, 311 Gates Hall, Ithaca, NY 14853-7501, USA
e-mail: djames@cs.cornell.edu

K. Anjyo (ed.), *Mathematical Progress in Expressive Image Synthesis I*,
Mathematics for Industry 4, DOI: 10.1007/978-4-431-55007-5_5,

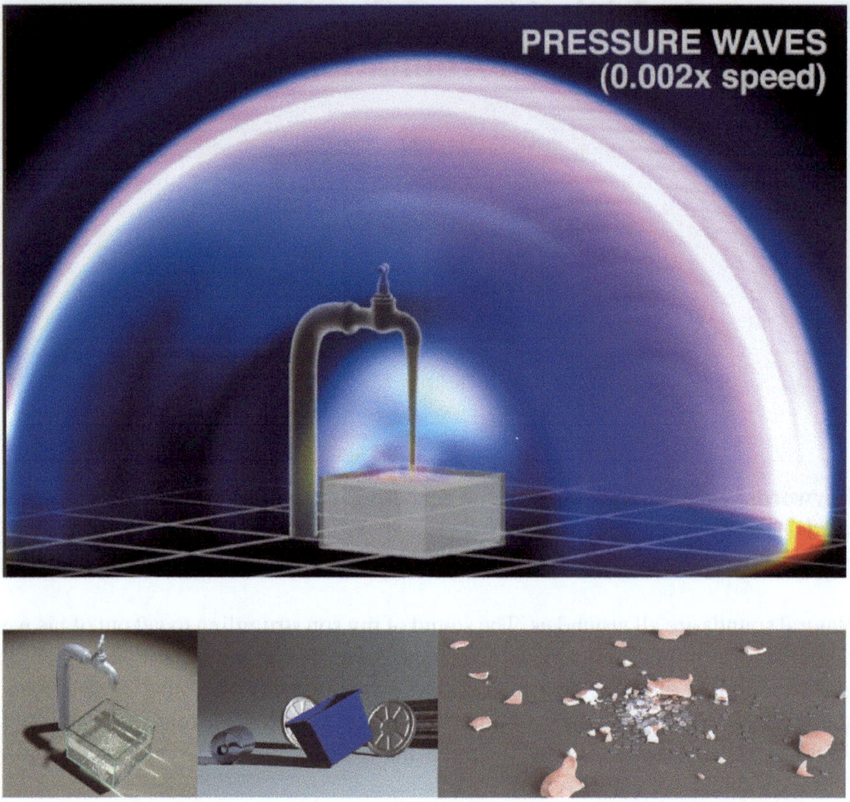

Fig. 1 Representative images from [2, 6, 7]

flexible bodies [8], liquids [6], thin-shell solids [2], brittle fracture [7], fire [3], and clothed virtual characters [1]. I will also highlight progress on mathematical methods and computer algorithms for reduced-order vibration analysis (linear and nonlinear), wave-based radiation analysis, precomputation techniques, reduced-order collision processing, many-body sound problems, sound propagation and listening, and real-time rendering. No prior knowledge of sound rendering will be assumed (Fig. 1).

References

1. An SS, James DL, Marschner S (2012) Motion-driven concatenative synthesis of cloth sounds. ACM Trans Graph 31(4):102:1–102:10.
2. Chadwick JN, An SS, James DL (2009) Harmonic shells: a practical nonlinear sound model for near-rigid thin shells. ACM Trans Graph 28(5):119:1–119:10.
3. Chadwick JN, James DL (2011) Animating fire with sound. ACM Trans Graph 30(4):84:1–84:8.

4. Chadwick JN, Zheng C, James DL (2012) Precomputed acceleration noise for improved rigid-body sound. ACM Trans Graph 31(4);103:1–103:9.
5. James DL, Barbic J, Pai DK (2006) Precomputed acoustic transfer: output-sensitive, accurate sound generation for geometrically complex vibration sources. ACM Trans Graph 25(3):987–995
6. Zheng C, James DL (2009) Harmonic fluids. ACM Trans Graph 28(3):37:1–37:12.
7. Zheng C, James DL (2010) Rigid-body fracture sound with precomputed soundbanks. ACM Trans Graph 29(4):69:1–69:13.
8. Zheng C, James DL (2011) Toward high-quality modal contact sound. ACM Trans Graph 30(4):38:1–38:12.

Efficient Image-Based Rendering Method Using Spherical Gaussian

Kei Iwasaki

Abstract Image-based rendering methods have been widely used to render realistic images of scenes illuminated by real-world, complex lighting. Recent advances on image-based rendering methods have revealed that spherical Gaussian functions have several nice properties for rendering and are suited to represent all-frequency signals compactly and accurately such as complex environment lighting and highly glossy BRDFs. This chapter introduces spherical Gaussians for efficient image-based rendering methods.

Keywords Image-based lighting · Real-time rendering · Spherical Gaussian

1 Introduction

Image-based rendering methods that capture the surrounding environment, store it as an environment map, and use it as the incident lighting (i.e. environment lighting) have been used to render realistic images. Real-time rendering of scenes illuminated by environmental lighting, is beneficial in many applications such as lighting/material design, animation, and games. For photo-realistic rendering under environmental lighting, the triple product of the environmental lighting, the BRDF, and the visibility function is integrated. Integrating the triple product is computationally expensive and this prevents from real-time rendering under all-frequency environment lighting. To address this, we propose an efficient rendering method for scenes illuminated by all-frequency environmental lighting. Our method uses spherical Gaussian (SG) functions to efficiently integrate the triple product.

K. Iwasaki (✉)
Wakayama University/JST CREST, 930, Sakaedani, Wakayama 640-8510, Japan
e-mail: iwasaki@sys.wakayama-u.ac.jp

K. Anjyo (ed.), *Mathematical Progress in Expressive Image Synthesis I*, Mathematics for Industry 4, DOI: 10.1007/978-4-431-55007-5_6,

2 Spherical Gaussian

A spherical Gaussian (SG) is a function over the unit sphere and a type of spherical radial basis functions. A spherical Gaussian (SG) $G(\boldsymbol{\omega}_i; \boldsymbol{\xi}, \eta, \mu)$ is represented by the following equation:

$$G(\boldsymbol{\omega}_i; \boldsymbol{\xi}, \eta, \mu) = \mu \exp(\eta(\boldsymbol{\omega}_i \cdot \boldsymbol{\xi} - 1)), \tag{1}$$

where the unit vector $\boldsymbol{\xi}$ is the lobe axis, η is the lobe sharpness, and μ is the lobe amplitude that consists of RGB components.

SG has nice properties for rendering. Firstly, the product of two SGs is represented by another SG. That is, the product of two SGs is closed in SG basis.

$$G(\boldsymbol{\omega}_i; \boldsymbol{\xi}_1, \eta_1, \mu_1) \cdot G(\boldsymbol{\omega}_i; \boldsymbol{\xi}_2, \eta_2, \mu_2) = G(\boldsymbol{\omega}_i; \boldsymbol{\xi}_3, \eta_3, \mu_3), \tag{2}$$

where the lobe axis ξ_3, the lobe sharpness η_3, and the lobe amplitude μ_3 are calculated by the following equations.

$$\xi_3 = \frac{\eta_1 \boldsymbol{\xi}_1 + \eta_2 \boldsymbol{\xi}_2}{||\eta_1 \boldsymbol{\xi}_1 + \eta_2 \boldsymbol{\xi}_2||}, \tag{3}$$

$$\eta_3 = ||\eta_1 \boldsymbol{\xi}_1 + \eta_2 \boldsymbol{\xi}_2|| \tag{4}$$

$$\mu_3 = \mu_1 \mu_2 \exp(||\eta_1 \boldsymbol{\xi}_1 + \eta_2 \boldsymbol{\xi}_2|| - \eta_1 - \eta_2) \tag{5}$$

Secondly, since SG is symmetric about the lobe axis, SG is easy to rotate just by rotating the lobe axis as the following equation.

$$\mathscr{R} G(\boldsymbol{\omega}_i; \boldsymbol{\xi}, \eta, \mu) = G(\boldsymbol{\omega}_i; \mathscr{R} \boldsymbol{\xi}, \eta, \mu), \tag{6}$$

where $\mathscr{R}$ is a rotation matrix.

Thirdly, the integral of SG over the unit sphere $\mathbb{S}^2$ is calculated analytically as the following equation.

$$\begin{aligned} \int_{\mathbb{S}^2} G(\boldsymbol{\omega}_i; \boldsymbol{\xi}, \eta, \mu) d\boldsymbol{\omega}_i &= \mu \int_0^{2\pi} \int_0^{\pi} \exp(\eta(\cos\theta - 1)) \sin\theta d\theta d\phi \\ &= \frac{2\pi\mu}{\eta}(1 - \exp(-2\eta)). \end{aligned}$$

Finally, the convolution of two SGs are calculated analytically as the following equation.

$$\int_{\mathbb{S}^2} G(\boldsymbol{\omega}_i; \boldsymbol{\xi}_1, \eta_1, \mu_1) \cdot G(\boldsymbol{\omega}_i; \boldsymbol{\xi}_2, \eta_2, \mu_2) d\boldsymbol{\omega}_i = \frac{4\pi\mu_1\mu_2}{\exp(\eta_1+\eta_2)} \frac{\sinh(||\eta_1\boldsymbol{\xi}_1+\eta_2\boldsymbol{\xi}_2||)}{||\eta_1\boldsymbol{\xi}_1+\eta_2\boldsymbol{\xi}_2||}.$$

Although the convolution of two SGs is calculated analytically as the above equation, it is not closed in SG basis. To address this problem, we have derived the SG representation of the convolution of two SGs [1].

$$\begin{aligned}
&\int_{\mathbb{S}^2} G(\boldsymbol{\omega}_i; \boldsymbol{\xi}_1, \eta_1, \mu_1) \cdot G(\boldsymbol{\omega}_i; \boldsymbol{\xi}_2, \eta_2, \mu_2) d\boldsymbol{\omega}_i \\
&\quad = \frac{4\pi\mu_1\mu_2}{\exp(\eta_1+\eta_2)} \frac{\sinh(||\eta_1\boldsymbol{\xi}_1+\eta_2\boldsymbol{\xi}_2||)}{||\eta_1\boldsymbol{\xi}_1+\eta_2\boldsymbol{\xi}_2||} \\
&\quad = 2\pi\mu_1\mu_2 \frac{\exp(||\boldsymbol{\zeta}||-\eta_1-\eta_2)-\exp(-||\boldsymbol{\zeta}||-\eta_1-\eta_2)}{||\boldsymbol{\zeta}||},
\end{aligned} \tag{7}$$

where $\boldsymbol{\zeta} = \eta_1\boldsymbol{\xi}_1 + \eta_2\boldsymbol{\xi}_2$. Here, we assume that both lobe sharpness values η_1 and η_2 are not too small. In practice, this assumption is valid for rendering applications. Then, we can assume that $e^{-||\zeta||-\eta_1-\eta_2} \approx 0$. Next, $||\boldsymbol{\zeta}||$ can be represented as follows:

$$\begin{aligned}
||\boldsymbol{\zeta}|| &= \sqrt{\eta_1^2||\boldsymbol{\xi}_1||^2 + 2\eta_1\eta_2\boldsymbol{\xi}_1\cdot\boldsymbol{\xi}_2 + \eta_2^2||\boldsymbol{\xi}_2||^2} \\
&= (\eta_1+\eta_2)\sqrt{1+\frac{2\eta_1\eta_2(\boldsymbol{\xi}_1\cdot\boldsymbol{\xi}_2-1)}{(\eta_1+\eta_2)^2}}.
\end{aligned} \tag{8}$$

By using this representation, the numerator in Eq. (7) can be calculated as follows:

$$\begin{aligned}
\exp(||\boldsymbol{\zeta}||-\eta_1-\eta_2) &= \exp((\eta_1+\eta_2)\left(\sqrt{1+\frac{2\eta_1\eta_2(\boldsymbol{\xi}_1\cdot\boldsymbol{\xi}_2-1)}{(\eta_1+\eta_2)^2}}-1\right)) \\
&\approx \exp((\eta_1+\eta_2)\cdot\frac{\eta_1\eta_2(\boldsymbol{\xi}_1\cdot\boldsymbol{\xi}_2-1)}{(\eta_1+\eta_2)^2}) \\
&= \exp\left(\frac{\eta_1\eta_2}{(\eta_1+\eta_2)}(\boldsymbol{\xi}_1\cdot\boldsymbol{\xi}_2-1)\right),
\end{aligned} \tag{9}$$

where a linear approximation of the Taylor expansion $\sqrt{1+x} \approx 1 + x/2$ is used. The denominator $||\boldsymbol{\zeta}||$ can be approximated with $\eta_1 + \eta_2$, since $\frac{2\eta_1\eta_2(\boldsymbol{\xi}_1\cdot\boldsymbol{\xi}_2-1)}{(\eta_1+\eta_2)^2}$ can be considered as negligible. Finally, the convolution of two SGs can be represented by a single SG as:

$$\int_{S^2} G(\boldsymbol{\omega}_i; \boldsymbol{\xi}_1, \eta_1, \mu_1) \cdot G(\boldsymbol{\omega}_i; \boldsymbol{\xi}_2, \eta_2, \mu_2) d\boldsymbol{\omega}_i \approx G\left(\boldsymbol{\xi}_1, \boldsymbol{\xi}_2, \frac{\eta_1\eta_2}{\eta_1+\eta_2}, \frac{2\pi\mu_1\mu_2}{\eta_1+\eta_2}\right). \tag{10}$$

3 Image-Based Rendering Using Spherical Gaussian

The outgoing radiance $L(\mathbf{x}, \boldsymbol{\omega}_o)$ at point $\mathbf{x}$ in the outgoing direction $\boldsymbol{\omega}_o$ under environment lighting is calculated by the following equation.

$$L(\mathbf{x}, \boldsymbol{\omega}_o) = \int_{\mathbb{S}^2} L(\boldsymbol{\omega}_i)V(\mathbf{x}, \boldsymbol{\omega}_i)\rho(\mathbf{x}, \boldsymbol{\omega}_i, \boldsymbol{\omega}_o)\max(0, \mathbf{n}(\mathbf{x})\cdot\boldsymbol{\omega}_i)d\boldsymbol{\omega}_i, \tag{11}$$

where $\mathbb{S}^2$ denotes the unit sphere in $\mathbb{R}^3$, $\boldsymbol{\omega}_i$ is the incident direction, $L(\boldsymbol{\omega}_i)$ is the distant lighting represented by the environment maps, $V(\mathbf{x}, \boldsymbol{\omega}_i)$ is the visibility function at $\mathbf{x}$, ρ is the BRDF, and $\mathbf{n}(\mathbf{x})$ is the normal at $\mathbf{x}$. To simplify the notation, we omit $\mathbf{x}$ in the following.Our method represents the BRDF ρ with $\rho(\boldsymbol{\omega}_i, \boldsymbol{\omega}_o) = k_d + k_s\rho_s(\boldsymbol{\omega}_i, \boldsymbol{\omega}_o)$ where k_d is a diffuse term and $k_s\rho_s(\boldsymbol{\omega}_i, \boldsymbol{\omega}_o)$ is a specular term. By substituting this in Eq. (11), $L(\boldsymbol{\omega}_o)$ can be calculated from the sum of the diffuse component L_d and the specular component $L_s(\boldsymbol{\omega}_o)$ as follows:

$$L(\boldsymbol{\omega}_o) = k_d L_d + k_s L_s(\boldsymbol{\omega}_o),$$

$$L_d = \int_{\mathbb{S}^2} L(\boldsymbol{\omega}_i)V(\boldsymbol{\omega}_i)\max(0, \boldsymbol{\omega}_i\cdot\mathbf{n})d\boldsymbol{\omega}_i, \tag{12}$$

$$L_s(\boldsymbol{\omega}_o) = \int_{\mathbb{S}^2} L(\boldsymbol{\omega}_i)V(\boldsymbol{\omega}_i)\rho_s(\boldsymbol{\omega}_i, \boldsymbol{\omega}_o)\max(0, \boldsymbol{\omega}_i\cdot\mathbf{n})d\boldsymbol{\omega}_i. \tag{13}$$

To calculate L_d, our method approximates $L(\boldsymbol{\omega}_i)$ and the cosine term $\max(0, \mathbf{n}\cdot\boldsymbol{\omega}_i)$ with the sum of spherical Gaussians.

Our method represents the environmental lighting L as the sum of SGs; $L(\boldsymbol{\omega}_i) \approx \sum_{k=1}^{K} G(\boldsymbol{\omega}_i; \boldsymbol{\xi}_k, \eta_k, \mu_k)$ where K is the number of SG lobes. The cosine term is also approximated by a single SG as $G(\boldsymbol{\omega}_i; \mathbf{n}, \eta_c, \mu_c)$. By substituting these terms into Eq. (12), L_d is calculated by the following equation.

$$L_d \approx \sum_{k=1}^{K} \int_{\mathbb{S}^2} G(\boldsymbol{\omega}_i; \boldsymbol{\xi}_k, \eta_k, \mu_k)G(\boldsymbol{\omega}_i; \mathbf{n}, \eta_c, \mu_c)V(\boldsymbol{\omega}_i)d\boldsymbol{\omega}_i. \tag{14}$$

The product of two SGs can be represented with a single SG:

$$G(\boldsymbol{\omega}_i; \boldsymbol{\xi}, \eta, \mu) = G(\boldsymbol{\omega}_i; \boldsymbol{\xi}_k, \eta_k, \mu_k)G(\boldsymbol{\omega}_i; \mathbf{n}, \eta_c, \mu_c), \tag{15}$$

where the lobe sharpness η is $||\eta_k\boldsymbol{\xi}_k + \eta_c\mathbf{n}||$, the lobe axis $\boldsymbol{\xi}$ is $(\eta_k\boldsymbol{\xi}_k + \eta_c\mathbf{n})/\eta$, and the amplitude μ is $\mu_k\mu_c e^{\eta-\eta_k-\eta_c}$. The diffuse component L_d is calculated by integrating the product of the visibility function $V(\boldsymbol{\omega}_i)$ and the spherical Gaussian $G(\boldsymbol{\omega}_i; \boldsymbol{\xi}, \eta, \mu)$. The integral of the product of SG and the visibility function is efficiently computed by using spherical signed distance function [3] for static scenes and by using integral spherical Gaussian [2] for dynamic scenes.

To calculate the specular component L_s, our method represents the BRDF ρ and the cosine term as the sum of SGs. As [3], the BRDF $\rho(\boldsymbol{\omega}_i, \boldsymbol{\omega}_o)$ can be calculated from the sum of the SGs as $\rho(\boldsymbol{\omega}_i, \boldsymbol{\omega}_o) \approx \sum_{j=1}^{J} G(\boldsymbol{\omega}_i; \boldsymbol{\xi}_j, \eta_j, \mu_j)$, where J is the number of SGs for ρ, and the lobe axis $\boldsymbol{\xi}_j$ depends on the outgoing direction $\boldsymbol{\omega}_o$. By substituting this into Eq. (13), $L_s(\boldsymbol{\omega}_o)$ can be rewritten as:

$$L_s(\boldsymbol{\omega}_o) \approx \sum_{j=1}^{J} \int_{\mathbb{S}^2} L(\boldsymbol{\omega}_i) G(\boldsymbol{\omega}_i; \boldsymbol{\xi}_j, \eta_j, \mu_j) G(\boldsymbol{\omega}_i; \mathbf{n}, \eta_c, \mu_c) V(\boldsymbol{\omega}_i) d\boldsymbol{\omega}_i. \tag{16}$$

Since the product of two SGs is also represented with a single SG $G_j(\boldsymbol{\omega}_i) = G(\boldsymbol{\omega}_i; \boldsymbol{\xi}'_j, \eta'_j, \mu'_j)$, $L_s(\boldsymbol{\omega}_o)$ can be rewritten as:

$$L_s(\boldsymbol{\omega}_o) \approx \sum_{j=1}^{J} \int_{\mathbb{S}^2} L(\boldsymbol{\omega}_i) G_j(\boldsymbol{\omega}_i) V(\boldsymbol{\omega}_i) d\boldsymbol{\omega}_i. \tag{17}$$

Our method approximates the integral of the triple product as:

$$L_s(\boldsymbol{\omega}_o) \approx \sum_{j=1}^{J} \int_{\mathbb{S}^2} L(\boldsymbol{\omega}_i) G_j(\boldsymbol{\omega}_i) d\boldsymbol{\omega}_i \frac{\int_{\mathbb{S}^2} G_j(\boldsymbol{\omega}_i) V(\boldsymbol{\omega}_i) d\boldsymbol{\omega}_i}{\int_{\mathbb{S}^2} G_j(\boldsymbol{\omega}_i) d\boldsymbol{\omega}_i}. \tag{18}$$

The numerator $\int_{\mathbb{S}^2} G_j(\boldsymbol{\omega}_i) V(\boldsymbol{\omega}_i) d\boldsymbol{\omega}_i$ can be calculated in the same way as L_d and the denominator $\int_{\mathbb{S}^2} G_j(\boldsymbol{\omega}_i) d\boldsymbol{\omega}_i$ can be calculated analytically as $\frac{2\pi}{\eta'_j}(1 - e^{-2\eta'_j})$. The integral of the product of L and G is calculated as a 3D function $\Gamma_L(\boldsymbol{\xi}, \eta)$:

$$\int_{\mathbb{S}^2} L(\boldsymbol{\omega}_i) G(\boldsymbol{\omega}_i; \boldsymbol{\xi}, \eta, \mu) d\boldsymbol{\omega}_i = \mu \Gamma_L(\boldsymbol{\xi}, \eta). \tag{19}$$

Our method precomputes $\Gamma_L(\boldsymbol{\xi}, \eta)$ and stores the data as prefiltered environment maps for various lobe sharpness η as [3]. Therefore, our method can change the BRDFs (i.e. the SGs) at run-time. Since our method integrates the product of the BRDF approximated by the SGs and the lighting, our method can represent highly specular materials.

Fig. 1 Rendering results of our method

4 Results and Conclusion

Figure 1 shows the rendering results of dynamic scenes illuminated by all-frequency environment lighting. We have implemented our rendering algorithm on a standard PC with an Intel Core i7 Extreme 965 CPU and a GeForce GTX 580 GPU. The rendering frame rates are 2.5–40 fps. The image resolutions are 640 × 480. We have proposed a real-time rendering method for fully dynamic scenes with highly specular BRDFs illuminated by all-frequency lighting by using spherical Gaussians.

References

1. Iwasaki K, Dobashi Y, Nishita T (2012) Interactive bi-scale editing of highly glossy materials. ACM Trans Graphics 26(6):144:1–144:7
2. Iwasaki K, Furuya W, Dobashi Y, Nishita T (2012) Real-time rendering of dynamic scenes under all-frequency lighting using integral spherical Gaussian. Comput Graphics Forum (Eurographics 2012) 31(2):727–734
3. Wang J, Ren P, Gong M, Snyder J, Guo B (2009) All-frequency rendering of dynamic spatially-varying reflectance. ACM Trans Graphics 28(5):133:1–133:10

A Lie Theoretic Proposal on Algorithms for the Spherical Harmonic Lighting

Masato Wakayama

Abstract The spherical harmonics are the angular portion of the solution to the Laplace equation in spherical coordinates and provide a frequency-basis for representing functions on the sphere. The spherical harmonic lighting, as defined by Robin Green at Sony Computer Entertainment in 2003, is a family of real-time rendering techniques that may produce certain realistic shading and shadowing with relatively small overhead lighting. All such spherical harmonic lighting techniques involve replacing parts of standard lighting equations with spherical functions that have been projected into a frequency space using the spherical harmonics as a basis (or a weight space of irreducible finite dimensional representation of the rotation group). In this chapter, using a group theoretical background of spherical harmonics and rather simple realization of the space of functions on the two dimensional sphere in the frame work of representation theory, we propose a possible geometry preserving algebraic/efficient computing, which might accelerate the (numerical and exact) computations slightly for spherical harmonic lighting.

Keywords Spherical harmonic lighting · Global illumination · Spherical harmonics · Irreducible representation · Intertwiner · Casimir element · Legendre polynomials

1 Introduction

The *spherical harmonics* are the angular portion of the solution to the Laplace equation in spherical coordinates. A first successful use of the spherical harmonics in computer graphics is found in [2], where they were used for interactive in-room

M. Wakayama (✉)
Institute of Mathematics for Industry, Kyushu University/JST CREST, 744, Motooka, Nishi-ku, Fukuoka 819-0395, Japan
e-mail: wakayama@imi.kyushu-u.ac.jp

K. Anjyo (ed.), *Mathematical Progress in Expressive Image Synthesis I*, Mathematics for Industry 4, DOI: 10.1007/978-4-431-55007-5_7,

lighting design. The *spherical harmonic lighting*, as defined by Robin Green at Sony Computer Entertainment in 2003 [4], is a family of real-time rendering techniques that may produce certain realistic shading and shadowing with relatively small overhead lighting. All such spherical harmonic lighting techniques involve replacing parts of standard lighting equations with spherical functions that have been projected into a frequency space (or, in our terminology, a weight space of irreducible finite dimensional representation of rotation group) using the spherical harmonics as a basis.

In this chapter, we first recall the basic notion/definition for representation theory briefly and provide a group theoretical background of spherical harmonics, and using this, we propose a possible geometry preserving algebraic/efficient computing, which might accelerate the (numerical and exact) computations slightly for spherical harmonic lighting [1, 4, 6, 7, 9] (see also [8, 10]) in some context. A mathematical idea presented here, if it actually works in the rendering process in computer graphics, would not be necessarily limited to the study of computer graphics, whence could be applicable to other fields. Our proposal is based on Lie theory or Representation theory of Lie groups. One of the general ideas or reason why this theory can work for such applications is, practically, to providing *simultaneous block-diagonalization* of matrices.

Spherical harmonics are orthogonal functions and span rotation invariant spaces on the two dimensional sphere S^2, allowing for efficient, alias-free least squares projection and reconstruction of spherical functions (= functions on the sphere). These properties lead to a number of efficient operations for computing rotations, convolution, and double product integrals [6, 9]. As is well-known, spherical harmonics are used extensively in various fields. They are a basis of the space L^2 (S^2) of the square integrable functions on S^2, as the name would suggest. They have been used to solve problems in physics, such as in heat equations, the gravitational and electric fields. They have also been used in quantum chemistry and physics to model the electron configuration in atoms. For the spherical harmonic lighting, in place of the Fourier series expansion on the Euclidean space, one uses the expansion by the spherical harmonics, in other words, replaces exponential functions by the associated Legendre functions (or Legendre spherical functions). From our current point of view, the (usual) Fourier analysis is considered to be based on very simple representation theory of abelian groups $\mathbb{R}^n$, whereas the spherical harmonics is on the representation theory of a non-commutative group $SO(3)$, $SO(n)$ being the rotation group of order n.

To be more explicit, we shall describe certain algebraic treatment for the computation of harmonic expansions, which turns to be a part of the technique at the spherical harmonic lighting, by the framework of harmonic analysis on the sphere $S^2 \cong SO(2)\backslash SO(3)$. More precisely, one considers the *irreducible decomposition* of the natural action defined by the right translation of $SO(3)$ on $L^2(S^2)$ (i.e. a part of the theory of spherical harmonics) and translate/reformulate the problem into the different Hilbert space using another but *unitarily equivalent realization* of irreducible representations on the space of polynomials with complex coefficients by the language of the special unitary group $SU(2)$ of degree two (see e.g. [3, 5, 11, 12]).

Here the word "unitarily equivalent" means the isometry between two Hilbert spaces with equivalent group actions.

2 Basic Notion for Representation Theory

One recalls here some of fundamental definitions and facts for representation theory of topological groups or Lie groups (see e.g. [3, 5, 11]: these text books also provide basic facts and terminologies about groups, Hilbert spaces, convergence of series, etc.). The readers may assume that Lie groups considered in this chapter are given by matrices groups such as $SO(n)$, the orthogonal group $O(n)$, the unitary group $U(n)$, the special linear group $SL_n(\mathbb{C})$, etc.

Definition 1 A unitary representation of a topological group G is a strongly continuous homomorphism π of G into the group $U(H)$ of unitary operators on a Hilbert space H. Here, a mapping $\pi : G \ni g \mapsto \pi(g) \in U(H)$ is called a homomorphism if it satisfies

$$\pi(gh) = \pi(g)\pi(h) \quad (\forall g, h \in G),$$

and a homomorphism π is called strongly continuous if the mapping $g \mapsto \pi(g)x$ is a continuous mapping of G into H for all $x \in H$. Moreover, one sometimes denotes the representation by a pair of the mapping π and representation space H as (π, H).

Definition 2 Two unitary representations (π_1, H_1) and (π_2, H_2) are called equivalent if there exists an isometry (i.e. a bijective linear mapping preserving the norm) A of H_1 onto H_2 satisfying

$$A\pi_1(g) = \pi_2(g)A \quad (\forall g \in G).$$

In this case, one writes $(\pi_1, H_1) \cong (\pi_2, H_2)$ (or simply $H_1 \cong H_2$). Obviously, the relation "$\cong$" is an equivalence relation.

Definition 3 Let (π, H) be a unitary representation of a group G. A closed linear subspace V of H is called invariant under π if one has

$$\pi(g)V \subset H \quad (\forall g \in G).$$

A unitary representation π is called irreducible if $H \neq \{0\}$ and H and $\{0\}$ are the only invariant subspaces of H.

Non-irreducible unitary representations are "decomposed" into irreducible representations. In a sense, the irreducible representations are the "atoms" of unitary representations.

Definition 4 Let (π_1, H_1) and (π_2, H_2) be two unitary representations of G. A linear map (not necessarily assuming an isomorphism) $A : H_1 \to H_2$ satisfying the relation in Definition 2, i.e. $A\pi_1(g) = \pi_2(g)A$ $(\forall g \in G)$, is called an *intertwiner* (or *intertwining operator*) between H_1 and H_2.

Proposition 1 *Let (π_1, H_1) and (π_2, H_2) be two finite dimensional unitary representations of G and A be an intertwiner between H_1 and H_2. Then, either $A = 0$ or A is a linear isomorphism.*

Proposition 2 *Let (π, H) be a unitary representations of G. Then, a closed subspace V of H is invariant under π if and only if the orthogonal projection P_V on V commutes with $\pi(g)$ for all $g \in G$. In this case, the orthogonal complement $V^\perp$ is also invariant under π.*

Let $\mathbf{B}(H)$ be the algebra of bounded linear operators on a Hilbert space H and M be a subset of $\mathbf{B}(H)$. The commutant M' of M is defined by

$$M' = \{L \in \mathbf{B}(H) \mid LU = UL \ (\forall U \in M)\}.$$

Theorem 1 (Schur's Lemma) *Let (π, H) be a unitary representations of G and $M = \{\pi(g) \mid g \in G\}$. Then, π is irreducible if and only if the commutant M' is equal to the set $\mathbb{C}1$ of scalar operators.*

Theorem 2 *Any unitary representation π of a compact group G is a (Hilbert space) direct sum of finite-dimensional irreducible unitary representations. In particular, any irreducible representation of a compact group is finite-dimensional.*

3 Spherical Harmonics

The groups treated in the present and subsequent sections such as the special unitary group $SU(2)$ (of degree 2) and special orthogonal groups $SO(2)$, $SO(3)$ are compact. Notice that since one always has an invariant inner product on the representation space by the existence of the Haar measure (invariant measure), any finite-dimensional representation of a compact group is assumed to be unitary [3, 11].

As a matrix group of degree 2, $SU(2)$ acts on the vector space $\mathbb{C}^2$: For $g = \begin{pmatrix} a & b \\ c & d \end{pmatrix} = \begin{pmatrix} \alpha & \beta \\ -\overline{\beta} & \overline{\alpha} \end{pmatrix} \in SU(2)$ $(|\alpha|^2 + |\beta|^2 = 1, \alpha, \beta \in \mathbb{C})$ the action is defined by

$$\mathbb{C}^2 \ni z = (z_1, z_2) \mapsto zg = (az_1 + cz_2, bz_1 + dz_2) \in \mathbb{C}^2.$$

Note that the action satisfies $z(g_1g_2) = (zg_1)g_2$ $(\forall g_1, g_2 \in SU(2))$ and $z1 = z$.

Let $\mathbb{C}[z_1, z_2]$ denote the polynomial algebra on $\mathbb{C}^2$. Let $V_m := \mathbb{C}[z_1, z_2]_m$ be the subspace of homogeneous polynomials of degree m of $\mathbb{C}[z_1, z_2]$. Then any polynomial f in V_m can be written uniquely as a linear combination of $m + 1$ monomials $z_1^k z_2^{m-k}$ $(0 \le k \le m)$. Hence one defines an $m + 1$ dimensional representation (π_m, V_m) of $SU(2)$ by

$$(\pi_m(g)f)(z) := f(zg).$$

The inner product defined by

$$(z_1^k z_2^{m-k},\ z_1^\ell z_2^{m-\ell}) = k!(m-k)!\delta_{k,\ell}$$

is invariant under π_m. In other words, equipped with this inner product, $(\pi_m,\ V_m)$ turns to be a unitary representation of $SU(2)$. One can prove the following theorem by the representation theory of Lie algebra $\mathfrak{sl}_2(\mathbb{C})$ of $SL_2(\mathbb{C})$.

Theorem 3 *For any non-negative integer m, the unitary representation* $(\pi_m,\ V_m)$ *of* $SU(2)$ *is irreducible. Moreover, any irreducible unitary representation of* $SU(2)$ *is equivalent to one of* $(\pi_m,\ V_m)$.

Theorem 4 *Let* $\mathbb{C}[z]_m$ *be the space of polynomials in z of degree less than or equals m. Then one defines the action* $\tau_m(g)$ *of* $g = \begin{pmatrix} \alpha & \beta \\ -\overline{\beta} & \overline{\alpha} \end{pmatrix} \in SU(2)$ *on* $\mathbb{C}[z]_m$ *by*

$$(\tau_m(g)p)(z) = (\overline{\beta}z + \alpha)^m p\left(\frac{\overline{\alpha}z - \overline{\beta}}{\overline{\beta}z + \alpha}\right) \quad (p \in \mathbb{C}[z]_m).$$

Equipped with the inner product defined by $(z^k,\ z^\ell)_m = \frac{k!(m-k)!}{(m+1)!}\delta_{k,\ell}$*, the representation* $(\tau_m,\ \mathbb{C}[z]_m)$ *is unitarily equivalent to* $(\pi_m,\ V_m)$.

Remark 1 The Hermitian inner product on $\mathbb{C}[z]_m$ defined in the theorem above can be expressed as follows:

$$(p_1,\ p_2)_m = \frac{m+1}{\pi}\int_{\mathbb{C}} p_1(z)\overline{p_2(z)}(1+|z|^2)^{-m-2}dz.$$

The representation theory of the 3-dimensional rotation group $SO(3)$ can be derived from that of $SU(2)$ described above, because $SU(2)$ is a (double) covering group of $SO(3)$. To see this, recall the adjoint representation Ad of $SU(2)$ defined by $\mathrm{Ad}(g)U = gUg^{-1}$ $(U, V \in \mathfrak{g},\ g \in SU(2))$ on the three dimensional real Lie algebra $\mathfrak{g} = \mathfrak{su}_2(\mathbb{R})$ (identified to the tangent space of $SU(2)$ at the identity 1 equipped with the Lie bracket $[U,\ V] := UV - VU$). Then, one easily observes that the kernel of Ad is given by $\{\pm 1\}$. From this

$$SU(2)/\{\pm 1\} \cong SO(3).$$

As a vector space, $\mathfrak{su}_2(\mathbb{R})$ is spanned by

$$X = \frac{1}{2}\begin{pmatrix} 0 & -i \\ -i & 0 \end{pmatrix},\quad Y = \frac{1}{2}\begin{pmatrix} 0 & -1 \\ 1 & 0 \end{pmatrix},\quad Z = \frac{1}{2}\begin{pmatrix} i & 0 \\ 0 & -i \end{pmatrix}.$$

Notice that $\{X, Y, Z\}$ is the orthogonal basis with respect to the inner product $\langle A, B\rangle := -2\mathrm{Tr}(AB)$ (the Killing form, see [3, 11]). It follows immediately that the map

$$\mathfrak{g} \ni xX + yY + zZ \mapsto \mathbf{x} = (x, y, z) \in \mathbb{R}^3$$

is an isometry. This fact actually gives the following.

Theorem 5 *For any non-negative integer ℓ, there exists an irreducible unitary representation ρ_ℓ of $SO(3)$ which is given by*

$$\rho_\ell \circ \mathrm{Ad} = \tau_{2\ell} (\cong \pi_{2\ell}).$$

Any irreducible unitary representation of $SO(3)$ is equivalent to ρ_ℓ for some ℓ. Moreover, if $\ell \neq m$, then ρ_ℓ is not equivalent to ρ_m.

Since the two dimensional sphere S^2 is realized by a homogeneous space (actually a compact Riemann symmetric space) of $SO(3)$ as $S^2 \cong SO(2)\backslash SO(3) \cong K\backslash SU(2)$, where $K := \{\begin{pmatrix} e^{i\theta/2} & 0 \\ 0 & e^{-i\theta/2} \end{pmatrix} \,|\, 0 \leq \theta < 4\pi\}$, one can naturally define a unitary representation of $SO(3)$ on the space of square integrable functions $L^2(S^2)$ on S^2 by right translation. Actually, we will see the irreducible unitary representations of $SO(3)$ are realized on the space of harmonic polynomials or spherical harmonics (thought as a well matched description).

Define a representation T_ℓ of $SO(3)$ on the space $\mathscr{P}_\ell$ of homogeneous polynomials of degree ℓ in three variables $\mathbf{x} = (x, y, z)$ by

$$(T_\ell(g)f)(\mathbf{x}) = f(\mathbf{x}g).$$

Notice that T_ℓ is not irreducible if $\ell \geq 2$, e.g. the space $\mathscr{P}_2$ contains the non-trivial invariant subspace spanned by the quadratic form $x^2 + y^2 + z^2$.

Let

$$\Delta := \frac{\partial^2}{\partial x^2} + \frac{\partial^2}{\partial y^2} + \frac{\partial^2}{\partial z^2}$$

be the Laplacian on $\mathbb{R}^3$. Define a space of harmonic polynomials of degree ℓ by

$$\mathscr{H}_\ell = \{f \in \mathscr{P}_\ell \,|\, \Delta f = 0\}.$$

Note that $\dim \mathscr{H}_\ell = 2\ell + 1$. Moreover, one knows the action of $SO(3)$ (the right translation T_ℓ) commutes with Δ, whence one can define a representation U_ℓ by

$$U_\ell = T_\ell|_{\mathscr{H}_\ell}.$$

Since any element f in $\mathscr{H}_\ell$ is a homogeneous polynomial of degree ℓ, for any $r \geq 0$ one has

$$f(rx, ry, rz) = r^\ell f(x, y, z).$$

Let $\mathscr{K}$ be the restriction of f to S^2. Put

$$\tilde{\mathscr{H}_\ell} := \mathscr{H}_\ell|_{S^2} = \{\mathscr{K}(f) \mid f \in \mathscr{H}_\ell\}.$$

Then the map $\mathscr{K}$ is a linear isomorphism of the vector space $\mathscr{H}_\ell$ onto $\tilde{\mathscr{H}_\ell}$. The space $\mathscr{H}_\ell$ is obviously recovered from $\tilde{\mathscr{H}_\ell}$ by the equation above. The elements of $\mathscr{H}_\ell$ are called spherical harmonics of degree ℓ. Since the space $\tilde{\mathscr{H}_\ell}$ is stable under the action $U_\ell(g)$ $(g \in SO(3))$, $(U_\ell, \tilde{\mathscr{H}_\ell})$ defines an irreducible unitary representation of $SO(3)$. Actually, one has the following.

Theorem 6 *As an irreducible unitary representation of* $SO(3)$

$$U_\ell \cong \rho_\ell.$$

The inner product on $\tilde{\mathscr{H}_\ell}$ is naturally given by

$$(f, g)_{S^2} = \int_{S^2} f(\mathbf{x})\overline{g(\mathbf{x})}d\mathbf{x},$$

where $d\mathbf{x}$ is the normalized measure on S^2 given by $d\mathbf{x} = (4\pi)^{-1} \sin\theta \, d\theta d\phi$. Here

$$\mathbf{x} = (x, y, z) = (r \sin\theta \cos\phi, r \sin\theta \sin\phi, r \cos\theta) \quad (r > 0, \ (\theta, \phi) \in [0, 2\pi] \times [0, \pi])$$

is the polar coordinates of $\mathbb{R}^3$.

Theorem 7 *The space* $\tilde{\mathscr{H}_\ell}$ *has an orthonormal basis* $\{Y_\ell^m\}_{-\ell \le m \le \ell}$ *defined by*

$$Y_\ell^m(\theta, \phi) = (-1)^m \sqrt{\frac{(2\ell + 1)(\ell + m)!}{(\ell - m)!}} e^{im\phi} P_\ell^{-m}(\cos\theta),$$

where $P_\ell^m(x)$ *is the associated Legendre function defined by*

$$P_\ell^m(x) = \frac{1}{2^\ell \ell!}(1 - x^2)^{m/2} \frac{d^{\ell+m}}{dx^{\ell+m}}(x^2 - 1)^\ell.$$

These Y_ℓ^m *are the matrix elements of the representation* $(U_\ell, \tilde{\mathscr{H}_\ell})$ *and called Legendre's spherical functions (see Theorem 8, Example 1).*

Remark 2 There is a relation:

$$P_\ell^{-m}(x) = (-1)^m \frac{(\ell - m)!}{(\ell + m)!} P_\ell^m(x).$$

Remark that several different normalizations are in common use (in quantum mechanics, seismology, geodesy, magnetics, etc.) for the Laplace spherical harmonic functions $Y_\ell^m(\theta, \phi)$.

A square integrable periodic function f on $\mathbb{R}$ (= a function on the torus $T := \mathbb{R}/\mathbb{Z}$) with the period 1 can be expressed as a Fourier series $f(x) = \sum_{n=0}^{\infty} a_n e^{2\pi inx}$ (L^2-convergence). Representation theoretically, this amounts to saying that $L^2(T)$ is decomposed as the direct sum of the irreducible (1-dimensional) unitary representations $\mathbb{R} \ni x \mapsto e^{2\pi inx} \in U(1)$ ($n \in \mathbb{Z}$) of $\mathbb{R}$ which is trivial on $\mathbb{Z}$. In the same way, since these representations U_ℓ exhaust all (equivalence classes) of irreducible unitary representations of $SO(3)$ one obtains

$$L^2(S^2) = \Sigma_{\ell=0}^{\infty} \oplus \mathscr{H}_\ell \quad \text{(a direct sum)}.$$

This shows that a (square integrable, hence in particular, continuous) function on the unit sphere S^2 can be expanded by the spherical harmonics $Y_\ell^m(\theta, \phi)$ $(-\ell \leq m \leq \ell,\ \ell = 0, 1, 2, \ldots)$. This fact is the consequence of the following Peter-Weyl theorem of compact group:

Theorem 8 *Let $\widehat{G} = \{\pi\}$ be the unitary dual of G, the set of all equivalence classes of irreducible unitary representations of G. Take a representative $(\pi,\ V)$ of π (using the same letter). Put $d_\pi = \dim_{\mathbb{C}} V$. Let $\{v_j\}_{1 \leq j \leq d}$ be the basis of V. Then the family $B_G := \{\sqrt{d_\pi}(\pi(g)v_i, v_j) \mid \pi \in \widehat{G},\ 1 \leq i, j \leq d_\pi\}$ is a complete orthonormal family of the Hilbert space $L^2(G)$ of all square integrable functions on G.*

Example 1 The unitary dual $\widehat{SO(3)}$ of $SO(3)$ can be parametrized by the set $\mathbb{Z}_{\geq 0}$ of non-negative integers ℓ. Since $S^2 \cong SO(2)\backslash SO(3)$, the Legendre spherical function is given by a matrix element $(\rho_\ell(g)f,\ f_0)$, where f_0 is an $SO(2)$-fixed vector, i.e. $\rho_\ell(k)f_0 = f_0$ for $k \in SO(2)$.

Remark 3 To illustrate the distributions of colors on the sphere determined by the harmonics basis, usually one replaces the basis $Y_\ell^m(\theta, \phi)$ by the real valued spherical harmonics $y_\ell^m(\theta, \phi)$ as follows.

$$y_\ell^m(\theta, \phi) := \begin{cases} \sqrt{2}\,\mathrm{Re}(Y_\ell^m)\ (m > 0) \\ Y_\ell^0\ (m = 0) \\ \sqrt{2}\,\mathrm{Im}(Y_\ell^m)\ (m < 0) \end{cases}.$$

The standard way to visualize the spherical harmonics can be seen in e.g. [6, 9]. For instance, one of those is to distort a unit sphere, by scaling each point radially by the absolute value of the function and soloing it based on the sign. Figure 1 illustrates such coloring only on the sphere for the first 5 spherical harmonics corresponding to the irreducible representation $(U_\ell,\ \tilde{\mathscr{H}_\ell})$ $(\ell = 0, 1, 2, 3, 4.)$ (Non-negative integers ℓ are called the bands). The green parts indicate regions where the function is positive, while white parts represent negative ones.

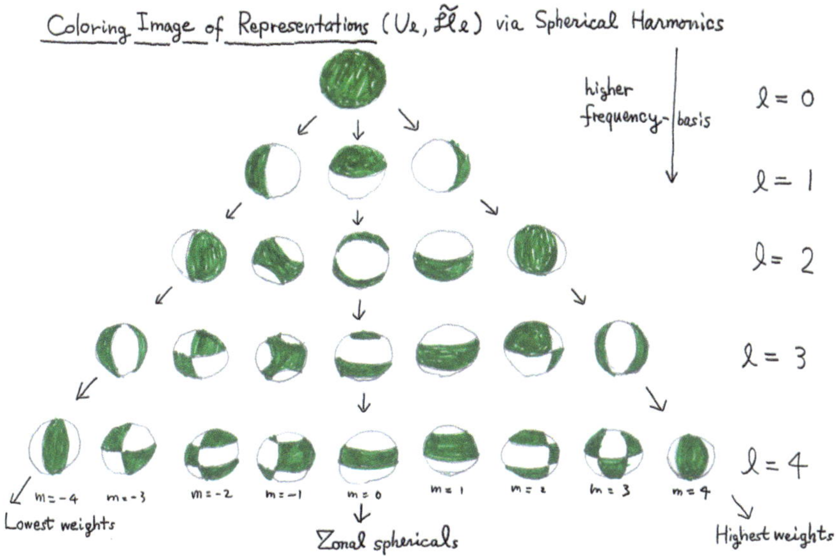

Fig. 1 Number $2\ell + 1$ of the spheres in each *horizontal row* is the dimension of the irreducible representation $(U_\ell, \mathscr{H}_\ell)$. (The figure is inspired from [6, 9])

4 Harmonic Expansions by Differentiation

Recall now the intertwiner A_ℓ (and its inverse) between $\mathbb{C}[w]_{2\ell}$ and $\mathscr{H}_\ell$. It is given explicitly as (see [3])

$$(A_\ell p)(\mathbf{x}) := \frac{2\ell + 1}{\pi} \int_{\mathbb{C}} p(w)\overline{H(\mathbf{x}, w)}^{\ell}(1 + |w|^2)^{-2\ell-2} dw,$$

where $H(\mathbf{x}, w) := (x + iy)w^2 + 2zw - (x - iy)$ (cf. Remark 1). Notice that, for w fixed, as a function of $\mathbf{x} = (x, y, z) \in \mathbb{R}^3$, one observes immediately that $H(\mathbf{x}, w)^\ell \in \mathscr{H}_\ell$.

By this description, one may transform the stage of calculations from the one using spherical harmonics to the one using simple monomials $z^m \in \mathbb{C}[z]_\ell$ by Theorem 5 together with Theorem 4. Namely, in some part of the spherical harmonic lighting technique, one might avoid rather complicated recurrence formulas and/or differential equation of the associated Legendre functions P_ℓ^m.

Recall the facts that in terms of the polar coordinate, Δ can be expressed as

$$\Delta = \frac{1}{r^2}\frac{\partial}{\partial r}\left(r^2 \frac{\partial}{\partial r}\right) + \frac{1}{r^2}\Delta_{S^2},$$

$$\Delta_{S^2} := \frac{1}{\sin\theta}\frac{\partial}{\partial\theta}\left(\sin\theta\frac{\partial}{\partial\theta}\right) + \frac{1}{\sin^2\theta}\frac{\partial^2}{\partial^2\phi}.$$

Notice that the last expression is actually known to be the image of the Casimir element $\mathscr{C} \in \mathscr{Z}U(\mathfrak{so}_3)$, the center of the universal enveloping algebra $U(\mathfrak{so}_3)(\cong U(\mathfrak{su}_2))$, under the (infinitesimal) right action by $SO(3)$. Therefore, since the representation U_ℓ is irreducible, one finds that $Y_\ell(\theta,\phi)$ is the eigenfunction of Δ_{S^2} with eigenvalue $-\ell(\ell+1)$ (by noticing that $\Delta r^\ell = \ell(\ell+1)r^\ell$):

$$\Delta_{S^2} Y_\ell(\theta,\phi) = -\ell(\ell+1)Y_\ell(\theta,\phi).$$

At the harmonic expansion of a spherical function f, one practically considers the approximation $\tilde{f}_N$ truncated by high frequency irreducible components U_ℓ ($\ell \geq N+1$):

$$f \approx \tilde{f}_N := \sum_{\ell=0}^{N}\sum_{|m|\leq\ell} a_\ell^m Y_\ell^m,$$

where we put $a_\ell^m = (f,\ Y_\ell^m)_{S^2}$ (but not computing here this integral). Let us consider the situation that one may assume that $f = \tilde{f}_N$ for some large N. (Actually, what one can detect practically is limited by bounded frequency components.)

Define the (projection) operator $P_\ell^N : \sum_{j=0}^N \oplus \mathscr{H}_j \to \mathscr{H}_\ell$ by

$$P_\ell^N := \prod_{\ell'=0,\,\ell'\neq\ell}^{N} \frac{\Delta_{S^2} + \ell'(\ell'+1)}{-\ell(\ell+1)+\ell'(\ell'+1)}.$$

It follows immediately that

$$P_\ell^N \tilde{f}_N = \sum_{|m|\leq\ell} a_\ell^m Y_\ell^m.$$

Then one has

$$a_\ell^m = (P_\ell^N \tilde{f}_N,\ Y_\ell^m)_{S^2} = (A_\ell^{-1} P_\ell^N \tilde{f}_N, z^{\ell-m})_{2\ell}.$$

Now we give an explanation how to obtain the inverse isomorphism A_ℓ^{-1} from the space $\mathscr{H}_\ell$ to $\mathbb{C}[w]_m$. Let $F \in \mathscr{H}_\ell$. Write F as

$$F(x,y,z) = g_0(x,y) + g_1(x,y)z + \cdots + g_\ell(x,y)z^\ell.$$

Note the fact that $g_j(x,y) \in \mathbb{C}[x,y]_{\ell-j}$ (i.e., is a polynomial of homogeneous degree $\ell - j$). Put $\Delta_{x,y} := \frac{\partial^2}{\partial x^2} + \frac{\partial^2}{\partial y^2}$. Then it is immediate that

$$\Delta_{x,y}F = \Delta_{x,y}g_0 + \cdots + \Delta_{x,y}g_\ell z^\ell,$$
$$\frac{\partial^2}{\partial z^2}F = 2g_2 + \cdots + \ell(\ell-1)g_\ell z^{\ell-2}.$$

Since g_ℓ (reps. $g_{\ell-1}$) is a constant (resp. a linear function) and $\Delta = \Delta_{x,y} + \frac{\partial^2}{\partial z^2}$, one observes that F is a harmonic polynomial if and only if the following condition are satisfied:

$$g_k = -\frac{1}{k(k-1)}\Delta_{x,y}g_{k-2} \quad (2 \le k \le \ell).$$

Hence one finds that there is a one-to-one correspondence between $\mathscr{H}_\ell$ and $\mathbb{C}[w]_{2\ell}$ as follows:

$$\begin{aligned}\mathscr{H}_\ell \ni F \mapsto (g_0, g_1) \in \mathbb{C}[x,y]_\ell \oplus \mathbb{C}[x,y]_{\ell-1} &\cong \mathbb{C}[w]_\ell \oplus \mathbb{C}[w]_{\ell-1} \\ &\cong \mathbb{C}[w^2]_\ell \oplus w\mathbb{C}[w^2]_{\ell-1} \cong \mathbb{C}[w]_{2\ell}\end{aligned}$$

Notice that the latter three isomorphisms are obviously all algebraic. Since $g_0 = F(x,y,0)$ and $g_1 = \frac{\partial}{\partial z}F(x,y,z)|_{z=0}$, using the reproducing kernel $K_z(w) = K(w,z) := (1+\bar{z}w)^{2\ell}$ (more precisely, the even and odd parts of the reproducing kernels) of the Hilbert space $\mathbb{C}[w]_{2\ell}$, one can essentially construct the inverse of the intertwiner A_ℓ.

This allows us to compute the coefficient a_ℓ^m from $P_\ell^N \tilde{f}_N$ avoiding numerical integration process such as using Monte-Carlo integration by random numbers. (via computing $P_\ell^N \tilde{f}_N(x,y,0)$ and $\frac{\partial}{\partial z}P_\ell^N \tilde{f}_N(x,y,z)|_{z=0}$. In other words, one may compute the inner product of the right hand side in purely algebraic way. Therefore, implementation of the idea provided in this section to computers for spherical harmonic lighting would be desirable.

References

1. Basri R, Jacobs DW (2003) Lambertian reflectance and linear subspaces. IEEE Trans Pattern Anal Mach Intell 25:218–233
2. Dobashi Y, Kaneda K, Nakashima E, Yamashita H, Nishita T (1995–1999) A quick rendering method using basis functions for interactive lighting design. Comput Graph Forum (Proc EUROGRAPHICS'95) 14(3):229–240
3. Faraut J (2008) Cambridge Studies in Advanced Mathematics. In: Analysis on lie groups, vol 110. Cambridge University Press, Cambridge
4. Green R (2003) Spherical harmonic lighting: gritty details, game developers conference 2003
5. Howe R, Chye TE (1992) Non-abelian harmonic analysis. Applications of $SL(2,\mathbb{R})$. Springer, New York
6. Schönefeld V (2005) Spherical harmonics. Seminal paper. http://videoarch1.s-inf.de/volker/prosem_paper.pdf
7. Seymour M (2013) The science of spherical harmonics at weta digital. http://www.fxguide.com/featured/the-science-of-spherical-harmonics-at-weta-digital/

8. Shirley P (2001) Realistic ray tracing. A K Peters, Natick
9. Sloan P-P (2008) Stupid spherical harmonics (SH) tricks. Game developers conference 2008
10. Sloan P-P, Kautz J, Snyder J (2002) Precomputed radiance transfer for real-time rendering in dynamic. Low-frequency lighting environments. Microsoft research and SIGGRAPH
11. Sugiura M (1975) Unitary representations and harmonic analysis. North-Holland/Kodansha, New York
12. Wakayama M (2013) Representation theory for digital image expression via spherical harmonics (in Japanese). In: Nishii R et al (eds) Mathematical approach to research problems of science and technology—theoretical basis and developments in mathematical modelling. MI lecture notes, vol 46. Kyushu University, Fukuoka

Interactive Editing of Volumetric Objects by Using Feature-Based Transfer Function

Yuhei Shibukawa, Yoshinori Dobashi and Tsuyoshi Yamamoto

Abstract This chapter focuses on the rendering of fluids represented by volumetric datasets. Generally, volume data can be visualized by a volume rendering technique. The result depends on the transfer function that converts a volume data into a colored volume. However, it is often difficult to create desired images by manually adjusting the transfer function. We propose an intuitive system that allows the user to directly design the appearance of the volume by specifying colors of the volume with a set of control points. The system solves an inverse problem to determine the transfer function such that the color of the rendered image becomes the same as those of the control points. In our method, the transfer function defined as a multi-dimensional function of features computed from the input volume dataset. Our method represents the transfer function as a linear combination of radial basis functions in order to efficiently solve the inverse problem.

Keywords Volume rendering · Fluid · Transfer function · Interactive editing

Y. Shibukawa (✉) · Y. Dobashi
Graduate School of Information Science and Technology,
Hokkaido University/JST CREST, Kita 14 Nishi 9, Kita-ku, Sapporo 060-0814, Japan
e-mail: shibukawa@ime.ist.hokudai.ac.jp

Y. Dobashi
e-mail: doba@ime.ist.hokudai.ac.jp

T. Yamamoto
Graduate School of Information Science and Technology, Hokkaido
University, Kita 14 Nishi 9, Kita-ku, Sapporo 060-0814, Japan
e-mail: yamamoto@ist.hokudai.ac.jp

K. Anjyo (ed.), *Mathematical Progress in Expressive Image Synthesis I*,
Mathematics for Industry 4, DOI: 10.1007/978-4-431-55007-5_8,

1 Introduction

Volume data represented on a three-dimensional grid is one of the more prevalent data structures in computer graphics. Volumetric objects, such as smoke and fire, are important elements that enhance the realism of synthetic scenes, and these are often represented by a type of volume data structure. Volume data can be visualized by a volume rendering technique. The result depends on the transfer function that converts a volume data into a colored volume. We focus on the design of the transfer function.

A transfer function is often used to assign the color and the opacity to each voxel. For example, a transfer function based on black body radiation is often used for rendering fire or explosions. Artists sometimes design their own transfer functions to produce the desired visual appearance of a volumetric object. However, this often requires a tedious trial and error process. In order to facilitate this process, we propose an interactive and intuitive system for editing the visual appearance by using a feature-based transfer function.

The method proposed in this chapter allows the user to edit the visual appearance of volumetric objects by setting control points to the rendered image directly. The user does not have to deal with a transfer function. While the user edits the visual appearance, our system inversely constructs the transfer function and the resulting images are rendered in real-time. Also, we propose the feature-based transfer function that computes the voxel properties from a set of features computed for each voxels. An important assumption behind our method is that artists expect similar visual appearances for regions with similar features. Our proposed system is achieved by using a feature-based transfer function that the user edits are reflected automatically in other regions having the same features. Examples of features include density, temperature, and their spatial derivatives. It is also possible to use geometric features such as the distance of each voxel above the bottom of the volume.

Our method represents the transfer function as a linear combination of radial basis functions (RBFs) in order to solve the inverse problem. The weights for the RBFs are computed by minimizing the difference between the edited image and the image of the volumetric object. The image of the volumetric object is then represented by a linear combination of basis images. Computing the transfer function is equivalent to finding the weights that minimize the difference between the edited image and the linear combination of basis images. This problem is represented by a simple matrix equation that can be solved very efficiently.

2 Related Work

In the field of scientific visualization, many methods have been proposed for designing transfer functions that can effectively visualize numerical simulations [2]. The methods used in this field enable transfer functions to be constructed so that important information can be properly visualized. However, these methods are not designed to edit the visual appearance of a volumetric object.

The user often needs to try different transfer functions repeatedly to create the desired visual appearance. Marks et al. [5] proposed a system that aids trial and error processes. This system creates a database of images by rendering the volume data many times with different randomly generated transfer functions and the user chooses one of them. However, it is not guaranteed that the desired result is always included in the database.

Wu and Qu proposed an interactive system for designing transfer functions by using genetic algorithms [6]. However, the computational cost of this method is too expensive to provide the user with real-time feedback. Kaneda et al. [1] proposed a method for rendering volume datasets efficiently by representing transfer functions with Fourier basis functions. This method allows us to render the volume efficiently even when the transfer function is modified. However, the user needs to try different transfer functions many times to obtain the desired result.

Some image processing techniques, such as colorization method [4], can be used to edit visual appearance of volumetric objects. In these methods, the user specifies the colors and opacities of a sparse set of voxels and the method automatically interpolates them for other voxels. However, it is not possible to directly edit the visual appearance on the rendered image and, in addition, the computational cost is too expensive for interactive editing.

The work related most closely to ours is the one recently proposed by Klehm et al. [3]. The method determines the voxel properties so that the image of a volumetric object becomes the same as the user-specified image. However, in Klehm's method, the visual appearance of the object viewed from a certain viewpoint is completely different from the visual appearances viewed from other viewpoints. Neither is the method suitable for dynamic volumetric objects. Furthermore, the method requires solving a large set of equations, resulting in an expensive computational cost. All the drawbacks of this method arise from the fact that the voxel properties are computed directly. Our method addresses these problems by introducing a feature-based transfer function.

3 Proposed Method

Figure 1 shows an overview of our system. The input to the system is a volume dataset and a precomputed radiance dataset. The volume dataset stores scalar values, such as the density and temperature, for each voxel. The precomputed radiance dataset is used to compute the intensity of each voxel in the volume. This dataset can be prepared using any of the previous methods, e.g., [7]. In computing the radiance dataset, we can take into account any lighting effects such as environmental illumination and multiple scattering. The user needs to specify a view point to render the volume before editing. Then the system computes a set of basis images in preprocess. Following the user's editing operations, our system constructs transfer functions by minimizing the energy

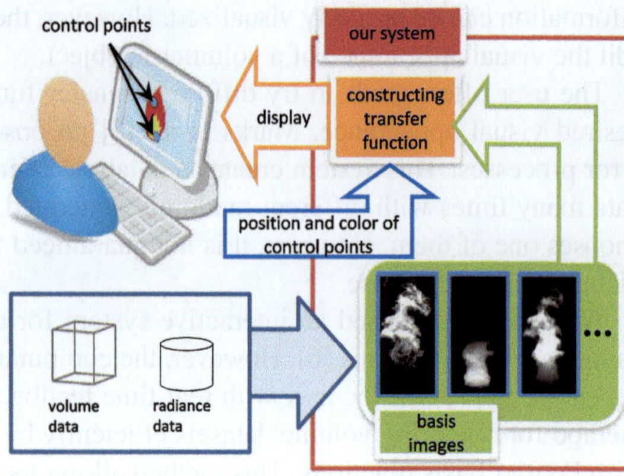

Fig. 1 This figure shows an outline of our system. By using our system, user can specifies desired color directly on the screen. Then our system computes transfer function by using basis images and display rendering result. Basis images are created from volume data and radians data in preprocess

functions that measure the differences between edited images and rendered images. The minimization problem is solved by using the basis images and the resulting image is displayed in real-time. Once a satisfactory visual appearance has been obtained, the user can verify the appearances by viewing from different viewpoints. For dynamic volumetric objects, the user can verify the visual appearance by observing different frames. Details of our method are described in the following subsections. In Sect. 3.1, we describe the volume rendering technique in our system. The basic idea in our method is to represent the transfer function by radial basis functions. The transfer function is determined by minimizing energy functions, and this is described in Sect. 3.2.

3.1 Volume Rendering

Our method is based on an equation that is often used for real-time volume rendering. The intensity of pixel p is expressed by the following equation.

$$I(p) = \int_{t_a}^{t_b} h(\mathbf{f}(\mathbf{x}(t))) I_v(\mathbf{x}(t)) g(\mathbf{x}_a, \mathbf{x}(t)) dt, \tag{1}$$

where h is a feature-based transfer function, $\mathbf{f}$ is a vector consisting of features of the volume at $\mathbf{x}$, and I_v is the intensity at point $\mathbf{x}$ toward the viewpoint. I_v is computed by using the precomputed radiance dataset. The transfer function h is introduced to modulate the intensity I_v.

$\mathbf{x}_a$ and $\mathbf{x}_b$ are the intersections between the viewing ray and the objects respectively (see Fig. 2). We assume that the light is attenuated exponentially, that is, g is given by:

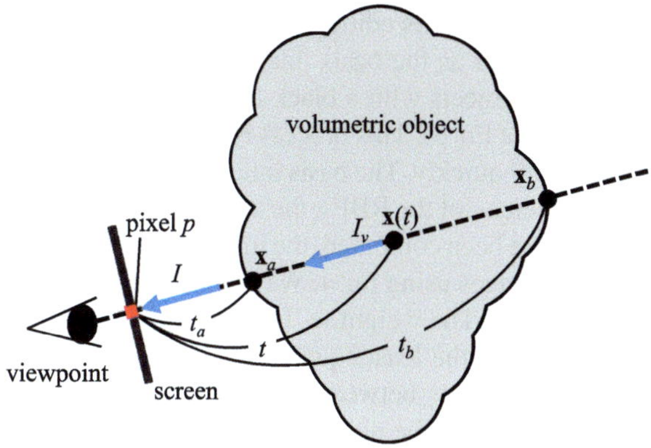

Fig. 2 This figure shows an overview of volume rendering technique. Our method is based on an equation that is often used for real-time volume rendering

$$g(\mathbf{x}_a, \mathbf{x}(t)) = \exp(-\kappa\tau(t_a, t)), \tag{2}$$

where κ is the attenuation ratio. $\tau(t_a, t) = \int_{t_a}^{t} \rho(s)ds$ is called the optical depth where ρ is the density at a point inside the object. The visual appearance of the volumetric object is designed by controlling κ and h.

3.2 Constructing the Transfer Function

The transfer function calculates intensity by future value of the volume at $\mathbf{x}$. The key idea in our method is to represent the transfer function h by using radial basis functions (RBFs), ϕ. That is,

$$h(\mathbf{f}(\mathbf{x})) = \sum_{i=0}^{n-1} w_i \phi(|\mathbf{f}(\mathbf{x}) - \mathbf{f}_i|), \tag{3}$$

where n is the number of RBFs, w_i is the weight for the i-th RBF and $\mathbf{f}_i$ is the center position of the i-th RBF in the multidimensional feature space. We use a Gaussian RBF: $\phi(r) = \exp(-cr^2)$, where c is a user-specified constant. By substituting the above equation into Eq. 1, we get:

$$I(p) = \sum_{i=0}^{n-1} w_i b_i(p), \tag{4}$$

$$b_i(p) = \int_{t_a}^{t_b} \phi(|\mathbf{f}(\mathbf{x}(t)) - \mathbf{f}_i|) I_v(\mathbf{x}(t)) g(\mathbf{x}_a, \mathbf{x}(t)) dt. \tag{5}$$

Before the user starts editing, our system computes $b_i(p)$ for each pixel using Eq. 5. $b_i(p)$ is stored as the basis image. The basis images are created by rendering the volumetric objects with a black background and by using ϕ as if it were a transfer function (see Eq. 5). This process is accelerated by the GPU and the basis images are created very quickly. The basis images need to be recomputed when the user changes the parameter c of the RBFs, the attenuation ratio κ, or the viewpoint. Once the basis images have been computed, the image I is efficiently created by the weighted sum of the basis images using Eq. 4. We do not have to repeat the volume rendering process to compute I. The weight w_i for each RBF is computed by minimizing the energy function using the least square method. The energy function, E, is defined by the sum of differences between the user-specified intensities and the intensities of the rendered image at the pixels $p \in Q$ (see Eq. 4). That is,

$$E = \sum_{p \in Q} \left(I_u(p) - \sum_{i=0}^{n-1} w_i b_i(p) \right)^2 + \sum_{i=0}^{n-1} w_i^2, \tag{6}$$

where I_u is the intensity specified by the user at pixel p using control point. The second term on the right is regularization term that is basically used to prevent overfitting and makes the transfer function as smooth as possible.

4 Results

This section shows several examples generated by our method. The examples shown in this section were computed on a desktop PC with an Intel Core i7 2600K (CPU) and an NVIDIA GeForce GTX560 (GPU). Figure 3 shows examples of the edited visual appearance of volumetric fire. The volume dataset stores the density and temperature for each voxel. A temperature and a distance of each voxel above the bottom of the volume were used as features. Figure 3a shows realistic fire that is created by the user. In Fig. 3b, the transfer functions determined for the Fig. 3a were interpolated for the other frames. In Fig. 3c, an artistic appearance has been created by making the fire reddish around the bottom and bluish around the top. Our method interpolates the user's specifications smoothly and naturally. Our method can be applied to various kind of volumetric objects. Figures 4 and 5 show examples of the edited visual appearance of volumetric clouds and explosion, respectively. The volume dataset of clouds stores the density for each voxel. The radiance dataset was prepared by taking into account the intensity due to direct sunlight and environmental illumination from a hemispherical light source with a uniform intensity distribution. These were stored separately. To edit the visual appearance, the density and environmental illumination were used. We prepared two different background images of the sky, and asked the user to design the visual appearance such that the clouds were naturally composited onto the background. Using our system, the user successfully designed such appearances as shown in the figure. The volume dataset of explosion consisted of density

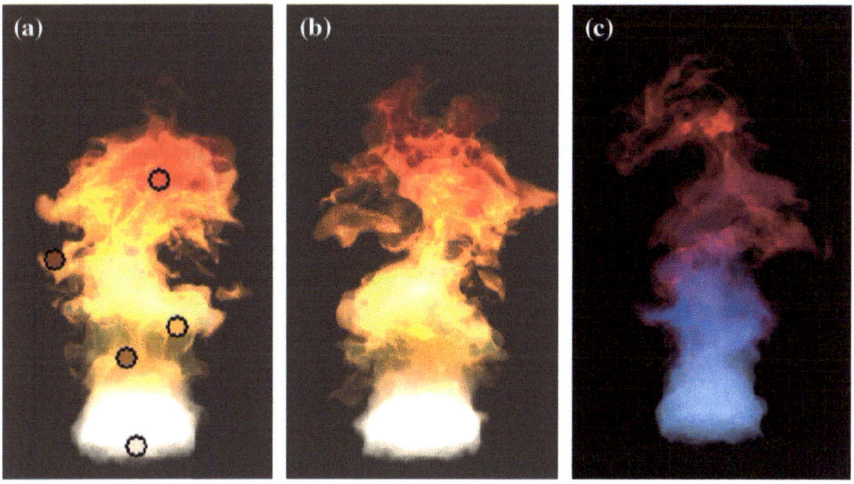

Fig. 3 Application of our method to volumetric fire. **a** realistic fire designed by the user, **b** the same fire as (a) rendered at a different frame, **c** artistic appearance designed by the user

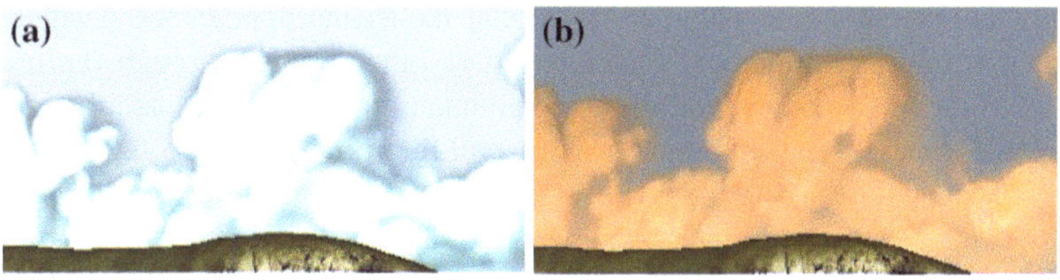

Fig. 4 Application of our method to volumetric cloud. **a** clouds at daytime, **b** clouds at sunset

Fig. 5 Application of our method to volumetric explosion

and temperature. The radiance dataset was precomputed by generating a set of point light sources for voxels with high temperature. In Fig. 5, we used the radiance for each voxel and the distance of each voxel above the bottom of the simulation space to edit the visual appearance. The user placed a set of control points to specify the colors as shown in the rightmost image, which corresponds to the last frame of the animation. Although the visual appearance of the last frame of the animation only was edited, similar appearances were generated for the other frames (the leftmost and center images).

5 Conclusions and Future Work

In this chapter, we have proposed a method for interactively designing the visual appearance of volumetric objects. A feature-based transfer function is introduced to generate various visual appearance of a volumetric object. Since we use features computed from the object, the visual appearance designed using an image rendered from a certain viewpoint is reproduced even if the objects is viewed from different viewpoints. For dynamic volumetric objects, the visual appearance designed for a certain frame is automatically transferred to other frames through the transfer function. Our method does not allow the attenuation ratio to be spatially variable. The spatially variable attenuation ratio is useful for volumetric objects consisting of multiple materials, e.g., smoke and fire. The user may want to adjust the amount of smoke or fire by assigning different colors and attenuation ratios. Our future work includes an extension of our method to editing such volumetric objects with multiple materials.

References

1. Kaneda K, Dobashi Y, Yamamoto K, Yamashita H (1996) Fast volume rendering with adjustable color maps. In: Proceedings of 1996 symposium on volume visualization, pp 7–14
2. Kindlmann G (2002) Transfer functions in direct volume rendering: design, interface, interaction. In: Proceedings of ACM SIGGRAPH course notes
3. Klehm O, Ihrke I, Seidel H, Eisemann E (2013) Volume stylize: tomography-based volume painting. In: Symposium on interactive 3D graphics and games, pp 161–167
4. Levin A, Lischinski D, Weiss Y (2004) Colorization using optimization. In: Proceedings of ACM SIGGRAPH 2004, pp 689–694, Aug 2004
5. Marks J, Mirtich B, Andalman B, Pfister H, Gibson G, Hodgins J, Kang T, Beardsley PA, Rum W, Freeman W (1997) Design galleries: a general approach to setting parameters for computer graphics and animation. In: Proceedings of ACM SIGGRAPH 1997, pp 389–400
6. Wu Y, Qu H (2007) Interactive transfer function design based on editing direct volume rendered images. IEEE Trans Visual Comput Graph 13(5):1027–1040
7. Zhou K, Ren Z, Lin S, Bao H, Guo B, Shum H (2008) Real-time smoke rendering using compensated ray marching. ACM Trans Graph 27(3) Article 36

Feature-Based Approach for the Interactive Editing of Environmental Lighting Effects

Munehiro Tada, Yoshinori Dobashi and Tsuyoshi Yamamoto

Abstract In computer graphics, it is not always guaranteed to generate user-desired shading effects by physically correct shading algorithms, due to the high computational cost and laborious work for parameter tuning. On the other hand, environmental map is a popular technique to create realistic shading images with low computational cost. Therefore, many methods have been proposed for editing shading effects obtained by environmental map. Since the methods edits the environmental map that represents only lights locating infinitely far away, these cannot edit local lighting effects, e.g., of spotlights. In this chapter, we propose an intuitive system that allows the user to produce the desired shading effects. Our system allows the user to specify desired-intensities at arbitrary positions on surfaces of objects, and design both local and global shading effects intuitively.

Keywords Shading · Interactive editing · Environmental lights · Radial basis function

1 Introduction

Recently, computer graphics techniques have been used in many applications such as movies, games and commercial films. One of the most important techniques for displaying realistic virtual objects is the ability to achieve appropriate shading. To

M. Tada (✉) · T. Yamamoto
Graduate School of Information Science and Technology, Hokkaido University,
Kita 14 Nishi 9, Kita-ku, Sapporo 060-0814, Japan
e-mail: tada@ime.ist.hokudai.ac.jp

T. Yamamoto
e-mail: yamamoto@ime.ist.hokudai.ac.jp

Y. Dobashi
Graduate School of Information Science and Technology, Hokkaido University/JST
CREST, Kita 14 Nishi 9, Kita-ku, Sapporo 060-0814, Japan
e-mail: doba@ime.ist.hokudai.ac.jp

K. Anjyo (ed.), *Mathematical Progress in Expressive Image Synthesis I*,
Mathematics for Industry 4, DOI: 10.1007/978-4-431-55007-5_9,

create photorealistic results, many physically-correct methods have been proposed [4, 12]. However, these physically correct methods require high computational cost, which makes the parameter tuning process time consuming to obtain desired results. To address this problem, image-based lighting (IBL) techniques have been proposed to create realistic results with reasonable computational cost [2]. These techniques employ environmental maps as virtual light sources [1]. Environmental maps are used to model real-world lighting environment and reproduce complicated shading effects caused by the natural illumination. One problem in these techniques is that the results are often different from the user-desired shading effects. Therefore, many methods have been proposed to address this problem [6, 7]. These methods allow the user to edit environmental maps intuitively to create user-desired shading effeces. Since the methods edits the environmental map that represents only lights locating infinitely far away, these cannot edit local lighting effects, e.g., of spotlights. In this chapter, we propose an editing system that allows the user to design both local and global shading effects intuitively.

Our system is based on feature-based interpolation of intensities and enables the user to edit shading effects obtained by environmental maps. In our system, the user directly specifies desired intensities on arbitrary positions of surfaces by using control points. Then, the shading effects are generated by feature-based interpolation of the user-specified intensities. We use features that are often used in the shading calculations, such as positions, normals, and visibilities, in order to produce plausible shading effects. By taking into account of positions as features for interpolation, the system achieves to edit shading effects locally. The interpolation function for the intensities is represented by a linear combination of radial basis functions (RBFs). The user can interactively add, move, or delete the control points, and the resulting shading effects are displayed in real time. Furthermore, our system allows the user to edit diffuse, and specular terms separately.

2 Previous Work

As we mentiond before, it is difficult to generate desired shading effects by using physically-correct shading models. Therefore, various methods have been proposed to edit shading effects even though the results are not physically-correct. Ritschel et al. proposed an editing method for specular reflection [8]. However, this method is limited to ideal specular reflections. Todo et al. proposed an editing method for cartoon shading effects [11]. However, this method is not suitable for photorealistic effects. Obert et al. proposed an editing method for shading effects by editing visibility functions [5]. Ritschel et al. proposed a method for deforming on-surface signals [9]. However these methods aim to deform shadow and shading, which makes difficult to specify user-desired intensities on arbitrary positions of the objects to be rendered. Methods for solving an inverse lighting problem for automatic placement of light sources has been proposed [3, 10]. However, these methods cannot reflect the user's intention on the resulting image.

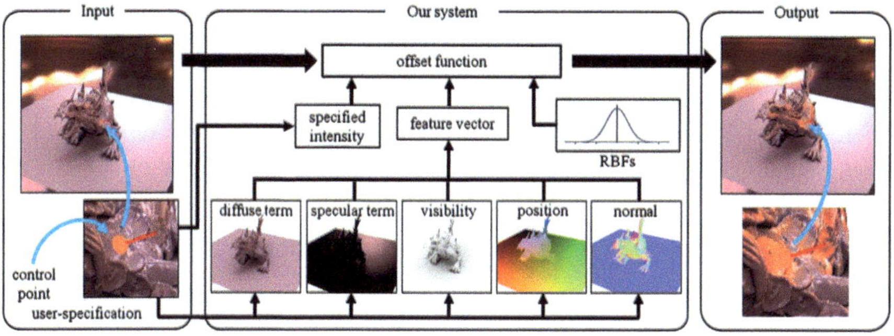

Fig. 1 Overview of proposed method

In this chapter, we propose an interactive editing system for shading effects obtained by environmental map. Okabe et al. proposed a method for calculating an environmental light by solving inverse problem that can produce specified shading effects [6]. However, it is not guaranteed that the specified shading effects are always achieved. Therefore, Pellacini proposed a method for modifying exiting environmental map [7]. However, in these methods, it is difficult to edit shading effects locally. Our system enables the user to edit shading effects locally.

3 Proposed Method

In our method, the intensity at point x on the object surfaces viewed from ω_o is computed by using the following equation.

$$I\left(x, \omega_o\right) = k_d\left(x\right) I_d\left(x\right) + k_s\left(x\right) I_s\left(x, \omega_o\right) \tag{1}$$

where I_d and I_s represent the intensities due to diffuse reflection and specular reflection, k_d and k_s are reflection coefficients for each reflection term respectively. Figure 1 shows an overview of our method. The user selects one of the two terms in Eq. (1) and places multiple control points on the surfaces of objects to specify the desired intensities at the positions of the control points. In Fig. 1, a single control point is placed as indicated by the orange sphere. The color of the sphere corresponds to the user-specified intensity. The red line indicates the normal direction at the control point. Our system then computes intensities on the surfaces of objects by using offset functions. Offset functions are determined by user-specified intensities and features that are often used in the shading calculation such as normal, position, and visibility. There are two offset functions corresponding to the two terms in Eq. (1). I_d and I_s are represented by:

$$\begin{aligned} I_d(x) &= \hat{I}_d(x) + h_d(x) \\ I_s(x, \omega_o) &= \hat{I}_s(x, \omega_o) + h_s(x, \omega_o) \end{aligned} \tag{2}$$

where h_d and h_s are the offset functions. The offset functions are represented by using RBFs. $\hat{I}_d$ and $\hat{I}_s$ are intensities before editing, and represented by:

$$\begin{aligned} \hat{I}_d(x) &= \int_\Omega L(\omega_i)\, V(x, \omega_i)\, (n_x, \omega_i)\, d\omega_i \\ \hat{I}_s(x, \omega_o) &= \int_\Omega L(\omega_i)\, V(x, \omega_i) \left(\omega_r(n_x, \omega_i), \omega_o\right)^s d\omega_i \end{aligned} \tag{3}$$

where Ω is a hemispherical domain, V is a visibility, ω_i is an incident direction, ω_r is a direction of reflection, n_x is a normal at point x, and s is a coefficient to determine the shininess of specular reflections. In our system, visibility is precomputed at each position. The user can interactively add, move, or delete the control points. During the user's operation, the offset functions are updated in real time and the resulting shading effects are displayed.

4 Offset Functions

4.1 *Constructing Offset Functions*

The offset functions, h_d and h_s are constructed by using the user-specified intensities. We will explain the construction of the offset function for the diffuse term, h_d, only. h_s is constructed in the same way. Let us assume that the number of control points is N, the user-specified intensity for each control point is $I_d^\star(x_i)$ $(i = 1, 2, \ldots, N)$, and the position of the i-th control point is x_i. The shading interpolator h_d is represented by:

$$h_d(x) = \sum_{i=1}^{N} a_i \phi\left(\|\mathbf{f}_d(x_i) - \mathbf{f}_d(x)\|\right) \tag{4}$$

where $\mathbf{f}_d(x_i)$ is a feature vector at x_i (see Sect. 4.2). a_i is a coefficient for each interpolation kernel ϕ, which is expressed by RBF. ϕ is given by:

$$\phi(r) = \exp\left(-r^2\right) \tag{5}$$

The Gaussian-type RBF is a continuous differentiable function and very popular as interpolation function. a_i is computed by using the least square method such that the following function is minimized:

$$\arg\min_{a_i} \sum_{j=1}^{N} \left(I_d^{\star}\left(x_j\right) - \left(\hat{I}_d\left(x_j\right) + h_d\left(x_j\right) \right) \right)^2 \tag{6}$$

where $\hat{I}_d\left(x_j\right)$ is diffuse component before editing. Our system computes a_i in real-time.

4.2 Feature Vectors

This section describes the features for the diffuse and specular terms. We employed each terms before editing, $\hat{I}_d\left(x\right)$ and $\hat{I}_s\left(x, \omega_o\right)$, as features to edit each term. In addition to $\hat{I}_d\left(x\right)$ and $\hat{I}_s\left(x, \omega_o\right)$, the coordinate p_x, the normal n_x, and visibility vector $\mathbf{V}\left(x\right) = \left(V\left(x, \omega_1\right), V\left(x, \omega_2\right), \ldots, V\left(x, \omega_n\right)\right)$ of the calculation point x are also included in the feature vector. In our system, the dimensions of $\hat{I}_d\left(x\right)$, $\hat{I}_s\left(x, \omega_o\right)$, the coordinate p_x, the normal n_x, and visibility vector $\mathbf{V}\left(\mathbf{x}\right)$ are 1, 1, 3, 3, and 32 respectively. Use of the visibility vector $\mathbf{V}\left(\mathbf{x}\right)$ enable the user to edit soft shadows. Use of the coordinate p_x allows the user to edit the shading effects locally. Use of the normal n_x makes it possible to control the shading effects according to the direction of surface. Consequently, the feature vectors, $\mathbf{f}_d\left(x\right)$ and $\mathbf{f}_s\left(x\right)$, are defined by:

$$\begin{aligned} \mathbf{f}_d\left(x\right) &= \left(\frac{\hat{I}_d\left(x\right)}{\sigma_d}, \frac{\mathbf{V}\left(x\right)}{\sigma_v}, \frac{p_x}{\sigma_p}, \frac{n_x}{\sigma_n} \right) \\ \mathbf{f}_s\left(x, \omega_o\right) &= \left(\frac{\hat{I}_s\left(x, \omega_o\right)}{\sigma_s}, \frac{\mathbf{V}\left(x\right)}{\sigma_v}, \frac{p_x}{\sigma_p}, \frac{n_x}{\sigma_n} \right), \end{aligned} \tag{7}$$

where $\sigma_d, \sigma_s, \sigma_p$ and σ_n are coefficients to control the effect of each component and are specified by the user. The user can enhance the effects of each component to reduce corresponding coefficient.

5 Results

This section shows results produced by using our method. We used a desktop PC with an Intel(R) Core(TM) i7-2600k 3.40GHz (CPU) and a NVIDIA GeForce GTX 560 (GPU). Our system is implemented on GPU and can interactively process at higher than 20 fps.

Figure 2 shows examples of the edited shading effects on a sappho (left column), a figurine (middle) and an armadillo. In Fig. 2, images in upper row show shading results before editing, and images in lower row show edited results. Figure 2d is an example of glossy specular effects like as metalic object. Figure 2e is an

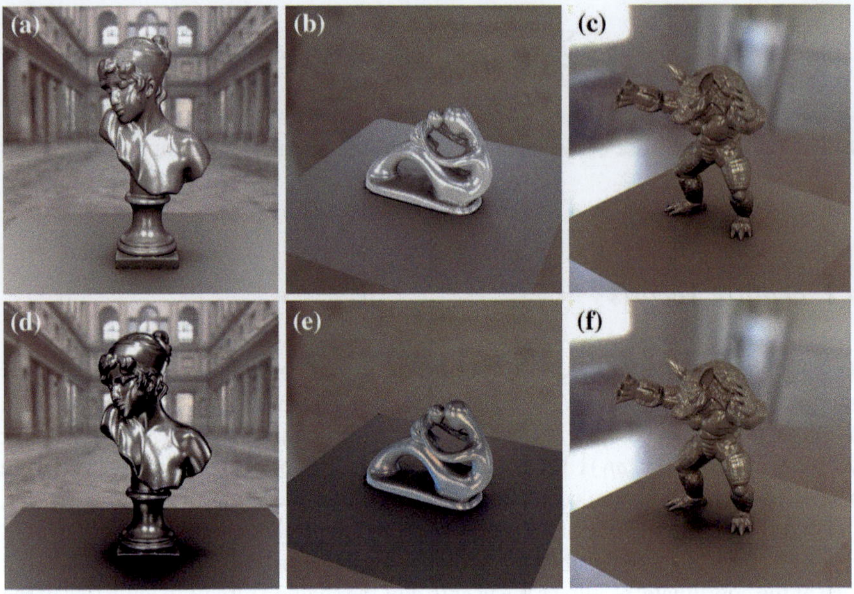

Fig. 2 Examples of edited *shading* effects. **a** Effects of glossy specular reflection (before editing), **b** Effects of skylight at sunset (before editing), **c** Effects of soft shadows (before editing), **d** Editing of glossy specular reflection shown in (a), **e** Editing of skylight effects shown in (b), **f** Editing of soft shadow effects shown in (c)

example of the effects of the skylight in sunset. Figure 2f is an example of the effects of soft shadows. Our system is useful for inserting synthetic 3D objects into real photographs, as shown in Fig. 3. The upper row in Fig. 3 shows rendered image by the Phong reflection model. By editing the shading of the synthetic objects (an asian doragon, a bunny and a space ship), the objects are naturally composited onto the real photographs as shown in lower images.

In general, the artificial shading effects shown in this section are often created by placing virtual light sources, but adjusting the parameters for such light sources is difficult and time-consuming. In contrast, our system allows the user to easily create the desired shading effects by directly specifying the desired color and intensity at the desired positions.

6 Conclusion and Future Work

In this chapter, we have proposed an interactive editing system for shading effects by feature-based interpolation. In our method, offset functions represented by RBFs are introduced. Our system allows a user to edit the shading effects for the diffuse, and the specular terms separately. By using our method, the user can intuitively design the desired shading effects.

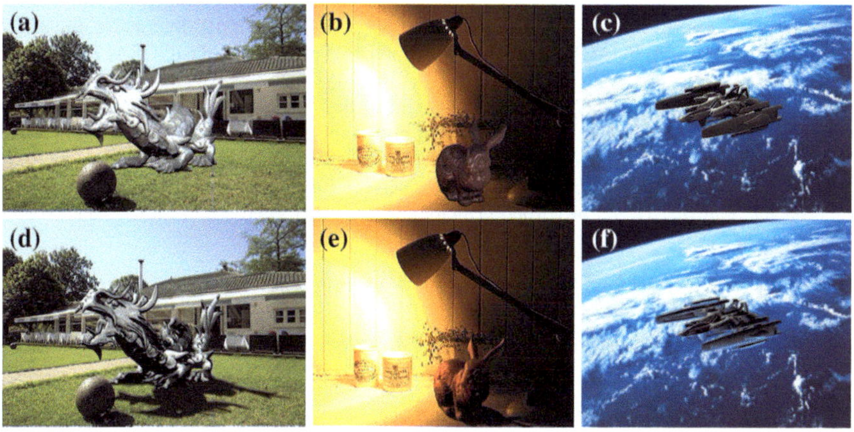

Fig. 3 Composition of synthetic objects into real photographs. **a** Insertion of a synthetic dragon into a photograph, **b** Insertion of a synthetic bunny into a photograph, **c** Insertion of a synthetic battleship into a photograph, **d** Editing of shading of the dragon, **e** Editing of shading of the bunny, **f** Editing of shading of the battleship

One of our future work is taking into account of interreflection effects. We are also interested in extending our system to other advanced shading models.

Acknowledgments This research is supported in part by Core Research for Evolutional Science and Technology (CREST) Program "Mathematics for Computer Graphics" of Japan Science and Technology Agency (JST).

References

1. Blinn JF, Newell ME (1976) Texture and reflection in computer generated images. Commun ACM 19(10):542–547
2. Debevec P (1998) Rendering synthetic objects into real scenes. In: Proceedings of SIGGRAPH, vol 98, pp 189–198
3. Gumhold S (2002) Maximum entropy light source placement. In: Proceedings of visualization 2002, pp 275–282
4. Jensen HW (1996) Global illumination using photon maps. In: Proceedings of the seventh eurographics workshop on rendering, pp 21–30
5. Obert J, Pellacini F, Pattanaik S (2010) Visibility editing for all-frequency shadow design. In: Computer graphics forum (Proceedings of eurographics symposium on rendering 2010), vol 29, No 4, pp 1441–1449
6. Okabe M, Matshshita Y, Shen L, Igarashi T (2007) Illumination brush: interactive design of all-frequency lighting. In: Proceedings of Pacific Graphics 2007, pp 171–180
7. Pellacini F (2010) envyLight: an interface for editing natural illumination. In: ACM transactions on graphics (Proceedings of SIGGRAPH 2010), vol 29, No 4
8. Ritschel T, Okabe M, Thormahlen T, Seidel HP (2009) Interactive reflection editing. In: ACM transactions on graphics (Proceedings of SIGGRAPH, Asia 2009), vol 28, No 5

9. Ritschel T, Thormahlen T, Dachsbacher C, Kautz J, Seidel HP (2010) Interactive on-surface signal deformation. In: ACM transactions on graphics (Proceedings of SIGGRAPH 2010) vol 29, No 4
10. Shacked R, Lischinski D (2001) Automatic lighting design using a perceptual quality metric. In: Computer graphics forum (Proceedings of Eurographics 2001), vol 20, NO 3, pp 215–227
11. Todo H, Anjyo K, Baxter W, Igarashi T (2007) Locally controllable stylized shading. In: ACM transactions on graphics (Proceedings of SIGGRAPH 2007), vol 26, No 3
12. Veach E, Guibas LJ (1997) Metropolis light transport. In: Proceedings of SIGGRAPH, vol 97, pp 65–76

Ray Tracing of Quadratic Parametric Surface

Shinji Ogaki

Abstract Over the past decades, vast research has been done on the ray-triangle intersect test but not much attention has been paid to the ray-quadratic parametric surface intersection test. In this chapter we present two direct ray tracing methods for quadratic parametric surfaces and introduce a simple optimization technique for them.

Keywords Ray tracing · Rendering · Parametric surface

1 Introduction

In the film industry there is an increasing demand for higher resolution images. The most common format has been Full HD (1920×1080 pixels, 2K) but 4K monitors or even 8K monitors are now available on the market. Today's computer synthesized objects are mostly modeled with polygons and their silhouettes often appear non-smooth when rendered in very high resolution. To obtain high quality images, models must be highly tessellated.

Ray tracing is becoming increasingly popular for photo-realistic image creation as it is a natural way to simulate the behavior of photons. Although a significant amount of research has been done to accelerate ray tracing, handling a large number of polygons is still costly. Memory consumption is another serious problem because the number of polygons is normally quadrupled as resolution is doubled. Directly performing ray tracing for parametric surfaces helps to reduce a required memory amount.

S. Ogaki (✉)
OLM Digital, Inc./JST CREST, Mikami Bldg. 2F, 1-18-10, Wakabayashi,
Setagaya-ku, Tokyo 154-0023, Japan
e-mail: shinji.ogaki@gmail.com

K. Anjyo (ed.), *Mathematical Progress in Expressive Image Synthesis I*,
Mathematics for Industry 4, DOI: 10.1007/978-4-431-55007-5_10,

However, we have to resort to numerical methods such as Bézier clipping [8] when the order of a parametric surface is high, we thus end up having heavy computational burden.

Our goal is to seek a sweet spot having a moderate amount of memory consumption and acceptable computation time. In what follows we focus on quadratic parametric surface since the intersection test is not so complicated as analytic solution exists.

2 Quadratic Parametric Surface

There are a variety of quadratic parametric surfaces including Steiner patch [3] and Phong tessellation [2]. In order to reduce memory consumption and achieve smooth rendering, we replace four triangle, one triangle and its adjacent three triangles, by a quadratic parametric surface.

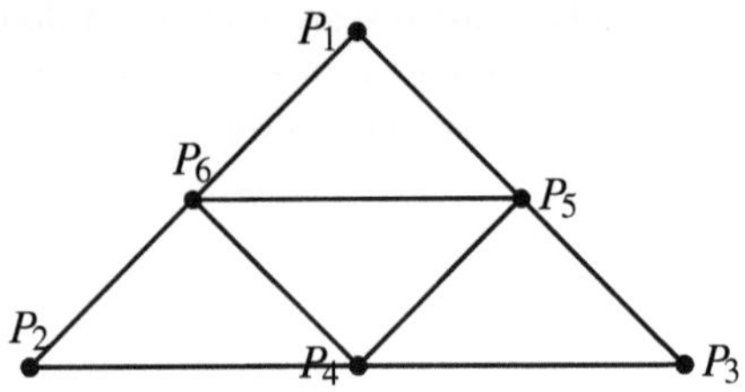

$$Q(u, v) = Au^2 + Bv^2 + Cw^2 + Duv + Euw + Fvw, \tag{1}$$

where $w = 1 - u - v$ and $0 \leq u, v, w \leq 1$. With the given six vertices P_1, P_2, P_3, P_4, P_5, and P_6 as in Fig. 2, the coefficients are determined as $A = P_1$, $B = P_2$, $C = P_3$, $D = 4P_6 - (P_1 + P_2)$, $E = 4P_5 - (P_1 + P_3)$, and $F = 4P_4 - (P_2 + P_3)$. This patch is C^0-continuous on edges.

3 Ray-Quadratic Parametric Surface Intersection Tests

There exist a number of methods for the ray-parametric surface intersection test including implicitization [4]. For higher order parametric surfaces we need numerical methods such as Newton's method and Beźier clippingBeźier clipping [8].

Here we start with describing the algorithm developed by Kajiya [6] because this method gives elegant solutions especially for the ray-quadratic parametric surface intersection test (Sects. 3.1 and 3.2).

In his method, each ray is represented as an intersection of two planes, say

$$0 = D_1 \cdot (x, y, z)^T + O_1 \tag{2}$$
$$0 = D_2 \cdot (x, y, z)^T + O_2. \tag{3}$$

The intersection of a ray and quadratic parametric surface lies on both the planes. Substituting the right-hand side of Eq. (1) for $(x, y, z)^T$ of Eqs. (2) and (3) gives two quadratic curves:

$$0 = F(u, v) = au^2 + bv^2 + c + duv + eu + fv \tag{4}$$

$$0 = G(u, v) = lu^2 + mv^2 + n + ouv + pu + qv, \tag{5}$$

where $a = (A + C - E) \cdot D_1$, $b = (B + C - F) \cdot D_1$, $c = C \cdot D_1 + O_1$, $d = (D-E-F+2C) \cdot D_1$, $e = (E-2C) \cdot D_1$, $f = (F-2C) \cdot D_1$, $l = (A+C-E) \cdot D_2$, $m = (B+C-F) \cdot D_2$, $n = C \cdot D_2 + O_2$, $o = (D-E-F+2C) \cdot D_2$, $p = (E-2C) \cdot D_2$, and $q = (F-2C) \cdot D_2$. Thus the intersection test for a bivariate quadratic parametric surface is rearranged into the intersection problem of two quadratic curves.

In the following subsections, we describe two interesting curve intersection finding algorithms.

3.1 The Method of Resultant

This method was introduced in [6]. Two quadratic curves have four intersections at a maximum and they are obtained analytically. We can find the intersections of the two planar curves $F(u, v)$ and $G(u, v)$ with the resultant by regarding them as quadratic curves of either u or v. If F and G have one or more common roots, the determinant of the Sylvester matrix is zero. If we treat the two curves as polynomials of u, we have

$$0 = a_4u^4 + a_3u^3 + a_2u^2 + a_1u + a_0, \tag{6}$$

where

$$\begin{aligned}
a_4 &= abo^2 + a^2m^2 + d^2lm + b^2l^2 - admo - bdlo - 2ablm \\
a_3 &= beo^2 + d^2mp \\
&\quad - admq - bdlq - bdop - afmo - demo - bflo \\
&\quad + 2(aem^2 + b^2lp + aboq + dflm - abmp - belm) \\
a_2 &= abq^2 + f^2lm + bco^2 + d^2mn + b^2p^2 + e^2m^2 \\
&\quad - bfop - bdno - efmo - cdmo \\
&\quad - bdpq - afmq - demq - bflq \\
&\quad + 2(b^2ln + acm^2 + beoq + dfmp - bemp - abmn - bclm) \\
a_1 &= beq^2 + f^2mp \\
&\quad - bfpq - bdnq - efmq - cdmq - bfno - cfmo \\
&\quad + 2(cem^2 + b^2np + bcoq + dfmn - bcmp - bemn) \\
a_0 &= bcq^2 + b^2n^2 + f^2mn + c^2m^2 - bfnq - cfmq - 2bcmn.
\end{aligned}$$

Solving this quartic equation, values for u are obtained. Substituting the values for u of Eq. (4) or (5), we obtain values for v. Unfortunately, this method is not robust because of numerical error.

3.2 The Method of Pencil

A simpler and more robust method is to utilize a matrix pencil [7]. The linear matrix pencil M is a linear combination of two matrices defined as

$$M = x \begin{pmatrix} a & d/2 & e/2 \\ d/2 & b & f/2 \\ e/2 & f/2 & c \end{pmatrix} + \begin{pmatrix} l & o/2 & p/2 \\ o/2 & m & q/2 \\ p/2 & q/2 & n \end{pmatrix} \tag{7}$$

with $x \in \mathbb{R}$. The linear combination of F and G can then be written as

$$0 = P = xF(u, v) + G(u, v) = (u, v, 1)M(u, v, 1)^T. \tag{8}$$

Interestingly, P can be represented as a product of two lines if $0 = |M|$. For more details, see [5] for example. This is a special case of a hyperbola. By letting

$$L_1 = \alpha_1 u + \beta_1 v + \gamma_1 \tag{9}$$

$$L_2 = \alpha_2 u + \beta_2 v + \gamma_2, \tag{10}$$

we have

$$(u, v, 1)M(u, v, 1)^T = L_1 L_2. \tag{11}$$

In this case the set of the intersections is decomposed as

$$\begin{aligned} \{(u, v)|0 = F\} \cap \{(u, v)|0 = G\} = \{(u, v)|0 = F\} \cap \{(u, v)|0 = L_1\} \\ \cup \{(u, v)|0 = F\} \cap \{(u, v)|0 = L_2\} \end{aligned}$$

since

$$\begin{aligned} \{(u, v)|0 = F\} \cap \{(u, v)|0 = G\} &= \{(u, v)|0 = F\} \cap \{(u, v)|0 = F + G\} \\ &= \{(u, v)|0 = F\} \cap \{(u, v)|0 = xF + G\} \\ &= \{(u, v)|0 = F\} \cap \{(u, v)|0 = P\} \\ &= \{(u, v)|0 = F\} \cap \{(u, v)|0 = L_1 L_2\}. \end{aligned}$$

Thus, the problem can be simplified by finding the value of x such that $0 = |M|$. The values of x are given by solving the following cubic equation

$$0 = a_3x^3 + a_2x^2 + a_1x + a_0, \tag{12}$$

where

$$\begin{aligned} a_3 =& abc + (def - af^2 - be^2 - cd^2)/4 \\ a_2 =& abn + amc + lbc - (afq + bep + cdo)/2 \\ &+ (oef + deq + dpf - lf^2 - me^2 - nd^2)/4 \\ a_1 =& amn + lbn + lmc - (lfq + mep + ndo)/2 \\ &+ (dpq + oeq + opf - aq^2 - bp^2 - co^2)/4 \\ a_0 =& lmn + (opq - lq^2 - mp^2 - no^2)/4. \end{aligned} \tag{13}$$

We only need to obtain one value of x, which significantly reduces the cost of root finding. The factorization is done by comparing the coefficients of u^2, v^2, uv, u, v, and 1. Dividing the both sides of Eq. (11) by M_{11} or M_{22}, whichever has a greater absolute value, makes the calculation easier and more robust.

In the case of $M_{11} > M_{22}$, the two lines L_1 and L_2 are obtained as

$$L_1 = u + \left(M'_{12} + \sqrt{M'^2_{12} - M'_{22}}\right) v + \left(M'_{13} \pm \sqrt{M'^2_{13} - M'_{33}}\right)$$

$$L_2 = u + \left(M'_{12} - \sqrt{M'^2_{12} - M'_{22}}\right) v + \left(M'_{13} \mp \sqrt{M'^2_{13} - M'_{33}}\right),$$

where $M'_{ij} = M_{ij}/M_{11}$. The coefficients γ_1 and γ_2 are chosen so that $M'_{23} = \beta_1\gamma_2 + \beta_2\gamma_1$.

Similary, in the case of $M_{22} > M_{11}$, L_1 and L_2 are obtained as

$$L_1 = \left(M'_{12} + \sqrt{M'^2_{12} - M'_{11}}\right) u + v + \left(M'_{23} \pm \sqrt{M'^2_{23} - M'_{33}}\right)$$

$$L_2 = \left(M'_{12} - \sqrt{M'^2_{12} - M'_{11}}\right) u + v + \left(M'_{23} \mp \sqrt{M'^2_{23} - M'_{33}}\right),$$

where $M'_{ij} = M'_{ij}/M_{22}$. The coefficients γ_1 and γ_2 are chosen so that $M'_{13} = \alpha_1\gamma_2 + \alpha_2\gamma_1$.

Substituting Eqs. (9) and (10) into Eq. (4) or (5) gives two quadratic equations. By solving them, we obtain values for u and v. These values give the actual intersection points. The parameters u, v, and $w = 1 - (u + v)$ must lie between 0 and 1, and intersection points must be in the viewing direction. Note also that the intersection test can be immediately terminated if $0 < M_{11}M_{22} - M_{12}M_{21}$ since $0 = (u, v, 1)M(u, v, 1)^T$ becomes an ellipsoid whose area is zero.

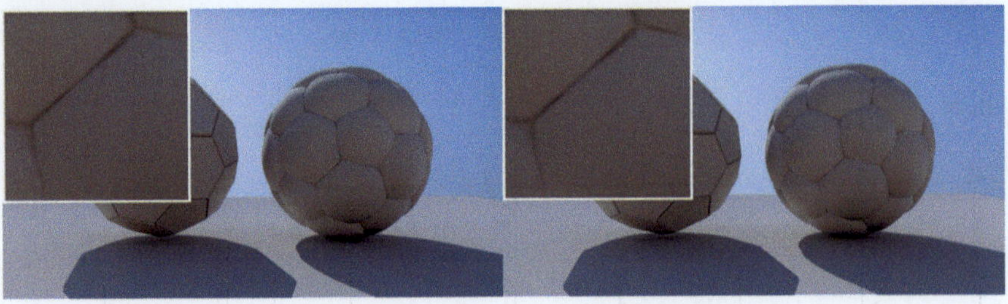

Fig. 1 Rendered images. The method of resultant (*left*) and method of pencil (*right*)

3.3 Optimization

In the previous section two intersection tests were introduced. What we have overlooked is the way of choosing the two planes (2) and (3). Here we show that the computation can be simplified if we choose them cleverly. For example, if D_1 and D_2 satisfy $a = 0$ and $m = 0$, the coefficients (13) become

$$\begin{aligned}
a_3 &= (def - be^2 - cd^2)/4 \\
a_2 &= lbc - (bep + cdo)/2 + (oef + deq + dpf - lf^2 - nd^2)/4 \\
a_1 &= lbn - (lfq + ndo)/2 + (dpq + oeq + opf - bp^2 - co^2)/4 \\
a_0 &= (opq - lq^2 - no^2)/4,
\end{aligned}$$

which dramatically reduces the number of multiplications.

4 Result

We applied Phong tessellation to a soccer ball model and rendered it with the two different algorithms as in Fig. 1. All calculations were done with single precision. A little artifact can be seen in the image rendered with the method of resultant. On the other hand, the method of pencil gives an ideal result.

We also rendered a simple quadratic parametric surface (Fig. 2) with three methods for performance comparison. The method of pencil with the optimization technique is 20 % faster than the method of resultant (Table 1).

5 Conclusion and Future Work

In this chapter we described two ray-quadratic parametric surface intersection tests. The method of pencil has a couple of advantages: (1) we can avoid to solve a quartic equation hence more robust results are obtained and (2) early termination of compu-

Fig. 2 A quadratic parametric surface

Table 1 Comparison of three methods

Resultant	Pencil	Pencil with optimization technique
1.00 s	0.84 s	0.80 s

tation is possible. We also showed that two well-chosen planes reduce the number of multiply operations. However, the intersection test still remains computationally expensive. We would like to explore a way to further improve the performance of the intersection test. Another interesting research avenue is to extend the pencil method for higher order parametric surfaces.

Acknowledgments We would like to thank Ken Anjyo and Sampei Hirose for their valuable comments. This work was partially supported by JST CREST.

Appendix

The geometric normal N_G of can be derived as

$$N_G = \left(\frac{\partial Q(u, v)}{\partial u} \times \frac{\partial Q(u, v)}{\partial v} \right) \Big/ \left| \frac{\partial Q(u, v)}{\partial u} \times \frac{\partial Q(u, v)}{\partial v} \right|. \tag{14}$$

The partial derivatives $\frac{\partial Q(u,v)}{\partial u}$ and $\frac{\partial Q(u,v)}{\partial v}$ are obtained as

$$\frac{\partial Q(u,v)}{\partial u} = (2A - E)u + (D - F)v + (E - 2C)w \tag{15}$$

$$\frac{\partial Q(u,v)}{\partial v} = (2B - F)v + (D - E)u + (F - 2C)w. \tag{16}$$

For smooth rendering, we use the Phong-interpolated normal N_p as in [3] instead of N_G. The geometric normal N_G should be used, for example, when the dot product of N_G and a reflected vector computed with N_p is negative.

References

1. Bajaj CL, Chen J, Xu G (1995) Modeling with cubic a-patches. ACM Trans Graph 14:103–133
2. Boubekeur T, Alexa M (2008) Phong tessellation. In SIGGRAPH Asia '08: ACM SIGGRAPH Asia 2008 papers, New York, NY, USA. ACM, pp 1–5
3. Breen DE (1986) Creation and smooth-shading of steiner patch tessellations. In: Proceedings of 1986 ACM fall joint computer conference, Los Alamitos, CA, USA, ACM '86. IEEE Computer Society Press, pp 931–940
4. Hanrahan P (1983) Ray tracing algebraic surfaces. In: Proceedings of the 10th annual conference on Computer graphics and interactive techniques, Detroit, Michigan, USA, SIGGRAPH '83. ACM, pp 83–90
5. Hosaka M (1992) Modeling of curves and surfaces in CAD/CAM. Springer, New York
6. Kajiya JT (1982) Ray tracing parametric patches. In: Proceedings of the 9th annual conference on Computer graphics and interactive techniques, New York, NY, USA, SIGGRAPH '82. ACM, pp 245–254
7. Ogaki S, Tokuyoshi Y (2011) Direct ray tracing of phong tessellation. In: EGSR' 11 Proceedings of the twenty-second eurographics conference on rendering, pp 1337–1344
8. Sederberg TW, Nishita T (1990) Curve intersection using bézier clipping. Comput Aided Des 22(9):538–549
9. Vlachos A, Peters J, Boyd C, Mitchell JL (2001) Curved pn triangles. In SI3D '01: Proceedings of the 2001 symposium on interactive 3D graphics, New York, NY, USA, 2001. ACM Press, pp 159–166

Part III
Fluid and Flow

A Flexible Image Processing Approach to the Surfacing of Particle-Based Fluid Animation (Invited Talk)

Ken Museth

Keywords Fluid animation · OpenVDB · FLIP · Particle skinning · Partial differential equations (PDE)

In recent years particle based techniques, like FLIP, have all but replaced Eulerian techniques for free-surface fluid animation in movie production. This, in turn, has put more emphasis on efficient tools that can turn point clouds into water surfaces. In this context one of the main challenges is to devise an approach that allows for both a high degree of artistic control as well as fast turnaround. In this chapter we outline such a system that has found good use at DreamWorks Animation. The core idea is surprisingly simple and yet powerful: create flexible and complex surface processing by means of daisy chaining simple and fast surface operators.

We have developed a novel approach to the important problem of turning particle systems into surfaces that represent an animated air-water interface. This system is open sourced in the C++ library OpenVDB [1], and has been thoroughly battle tested at DreamWorks Animation during the production of "The Croods" [2] (see Fig. 1) and "How to Train Your Dragon 2". A distinguishing feature of our system is the fact that it is based on a flexible combination of numerous fast surface processing operators, several of which are 3D generalizations of proven techniques from image processing. This greatly empowers artists to quickly customize and iterate in a more sequential and manageable manner, hence allowing a significant degree of artistic control which is paramount in any movie production. This is a departure from most existing surfacing systems that are based on more complex monolithic techniques that are relatively slow, typically employ non-intuitive parameters and offer limited artistic control. Examples of such monolithic turn-key systems that attempt to generate high-quality surfaces in a single computationally expensive step are techniques

K. Museth (✉)
DreamWorks Animation, Glendale, CA, USA
e-mail: Ken.Museth@DreamWorks.com

K. Anjyo (ed.), *Mathematical Progress in Expressive Image Synthesis I*, Mathematics for Industry 4, DOI: 10.1007/978-4-431-55007-5_11,

Fig. 1 Example of particle skinning for Lagrangian fluid animation using our framework based on VDB and a sequence of highly customizable and multithreaded surface operators. *Left* The input particles generated by a FLIP fluid simulator (Naiad). Note the low density of particles colliding with the character. *Right* The resulting liquid surface represented by an adaptive quad-mesh. Note the presence of thin sheets and sharp feature on an otherwise smooth surface with little signs of spatial aliasing due to particle undersampling. This image is property of DreamWorks Animation

employing elliptic particle footprints, initially proposed in [3, 4] and subsequently in [5], and techniques using surface filters based on the solution of high-order parabolic partial differential equations, e.g. [6]. While such techniques are impressive in their own right, they have proved to be undesirable in production where artists prefer a faster and more layered approach in which a specific "look and feel" is progressively achieved by daisy chaining simpler surface operators. Curiously this approach is similar in spirit to most existing image processing or 2D compositing workflows.

Our technical contributions are twofold; foremost, as outlined above, the simple yet surprisingly powerful idea to base our framework on a rich toolbox of fast low-level surface operators that can be combined in any order since they share a common implicit representation. Second, to the best of our knowledge several of these 3D operators are novel, though many of them have strong ties to conceptually similar ideas known from image processing.

The proposed workflow has three fundamental stages: initialization, surface processing and meshing. The first step simply serves to quickly convert the particles to a representation on which all the subsequent operations can work. This choice of representation is a narrow-band level set that can be represented compactly

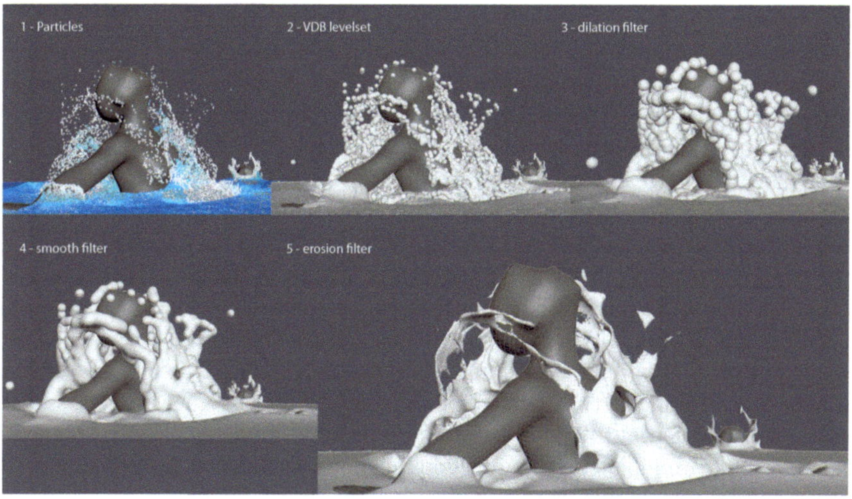

Fig. 2 Simplified illustration of a typical combination of surface processing operations for animated particle skinning: *1* Particles generated from any Lagrangian fluid simulation package. *2* Multithreaded conversion of particles into narrow-band level sets that merely serve as the initialization for our subsequent surface processing operators. *3* Dilation filter with interface tracking. *4* Application of various convolution filters with interface tracking, *5* Erosion filter with surface tracking. This image is property of DreamWorks Animation

with VDB [1], and offers excellent computational performance for complex surface deformations, which of course are a hallmark of liquid interfaces. The conversion from particles to signed distance fields is easily multithreaded since VDB supports efficient hierarchical CSG operations required when joining partial conversions performed by the different computational threads. To improve temporal coherence this stage also supports simple velocity stretching and attenuation of particle footprints to form teardrop shapes.

The next (main) step in our pipeline applies different types of surface operators, all based on the implicit level set representation. We performed this post-processing of the initial particle surface by means of various combinations of custom filtering and morphological operations. These surface processing operators can be grouped into three distinctive kinds. The first group of tools is based on morphological operators like dilation, erosion, closing and opening. They effectively allow the artist to fill holes, remove small isolated particles, and sharpen, peak or blur surface details. Next, are smoothing operators that are based on differential properties of the level set surface. This includes mean-curvature flow and Laplacian smoothing that perform second-order smoothing operations. Last but not least, we use various types of kernel-based convolution filters that can smooth (or sharpen) surface details in a very computationally efficient manner. Examples include Gaussian, mean-value and median-value 3D filters that are applied directly to the signed distance fields. Unlike the curvature-based smoothing that solves parabolic PDEs, the kernel-based filtering techniques facilitate much faster surface deformations while still allowing for proper interface tracking and re-normalization of the narrow-band. This approach is especially attractive when employing filters that can be represented as a sequence of separable convolution kernels, e.g. box filters. As a simple but essential modification, these surface operators can all be augmented by arbitrary user-defined alpha-masks. This greatly enhances their usefulness since it allows artists to localize and better control the effects of an operation, for instance by deriving alpha-masks from differential properties of surfaces (e.g. curvature) or fluid velocity fields (e.g. vorticity). Alternatively these masks can also be created by painting directly on surfaces. Another very useful type of alpha-mask constraints the surface deformations to a user defined proximity to the input particles. Essentially this allows for the use of aggressive filtering and smoothing while preserving small surface details like ballistic water droplets. This idea was first proposed for polygonal surface in [7], but as demonstrated in [6] it is easily reformulated to level set surfaces. Figure 2 shows a simple example of a common combination of operators corresponding to a morphological closing that is interlaced with kernel-based filtering.

The final stage of our operator pipeline converts the narrow-band level sets to a polygonal mesh. Since VDB allows for very high resolution volumes we typically employ adaptive meshing techniques (e.g. a modified dual-contouring scheme) to greatly reduce the polygon count.

References

1. Museth K (2013) VDB: high-resolution sparse volumes with dynamic topology. ACM Trans Graph 32(3):27:1–27:22. http://www.openvdb.org
2. Budsberg J, Losure M, Museth K, Baer M (2013) Liquids in "The Croods". ACM DigiPro.
3. Museth K, Clive M, Zafar NB (2007) Blobtacular: surfacing particle systems in "Pirates of the Caribbean 3". ACM SIGGRAPH sketches and applications, New York
4. Museth K (2012) System and method for surfacing of particle systems. Patent No. US 8,199,148 B2, filed by Digital Domain 1 August 2008, issued 12 June 2012.
5. Yu J, Truk G (2013) Reconstructing surfaces of particle-based fluids using anisotropic kernels. ACM Trans Graph 32(1):5:1–5:12.
6. Bhattacharya H, Gao Y, Bargteil AW (2011) A level-set method for skinning animated particle data. In: ACM SIGGRAPH/Eurographics symposium on computer animation.
7. Williams B (2008) Fluid surface reconstruction from particles. Master's thesis, University of British Columbia, Columbia.

Inverse Approach for Visual Simulation of Clouds

Yoshinori Dobashi

Abstract Clouds play an important role for creating realistic images of outdoor scenes. There are two important factors in synthesizing realistic images, that is, shapes and colors of clouds. Many methods have therefore been proposed for modeling and rendering clouds. One of the promising approaches is to numerically simulate the actual physical phenomena. However, realistic images cannot be generated unless the user chooses appropriate parameters involved in the numerical simulation, which is not an easy task. The shapes and colors of the simulated clouds depend on many parameters and it is generally difficult and time-consuming to adjust those parameters manually. This paper presents an inverse approach to address this problem. For cloud shapes, we present a method for controlling the simulation of cloud formation so that the simulated shapes become similar to those specified by the user. For colors of clouds, a method for automatically adjusting the parameters for computing realistic colors by using user-specified photographs of real clouds is presented.

Keywords Clouds · Inverse rendering · Fluid simulation · Genetic algorithm · Feedback control

1 Introduction

Clouds are important elements when synthesizing images of outdoor scenes to enhance realism. Many methods have therefore been proposed for visual simulation of clouds [1, 4, 6–10]. These methods are used in many applications such as flight simulators, movies, computer games, and so on. With the rapid development of the computers, recent researches focus on the numerical simulation of the actual physi-

Y. Dobashi (✉)
Graduate School of Information Science and Technology, Hokkaido University/JST CREST,
Kita 14 Nishi 9, Kita-ku, Sapporo 060-0814, Japan
e-mail: doba@ime.ist.hokudai.ac.jp

K. Anjyo (ed.), *Mathematical Progress in Expressive Image Synthesis I*,
Mathematics for Industry 4, DOI: 10.1007/978-4-431-55007-5_12,

cal phenomena. However, one of the problems is that it is often difficult to achieve desired appearances of the clouds. The appearances of the clouds are determined by the shapes and colors. Animators need to adjust many non-intuitive parameters manually by a trial-and-error process. The expensive computation cost of the numerical simulation makes this process much more difficult. One of the solutions to this problem is to accelerate the simulation process. This can be considered as a fast solution to the forward problem: the corresponding output image is computed in real-time using the given parameters, allowing one to efficiently find the appropriate parameters that produce a desired appearance. However, even using this approach, a repetitive trial-and-error process is still required until satisfactory results are obtained. Our aim is to remove the manual trial-and-error process by solving an inverse problem.

For the shapes of clouds, we present a method for controlling the simulation of the cloud formation process [5]. In this method, the user specifies the contours of the clouds viewed from a specific camera position. The simulation process is controlled in order to form the desired shape of the clouds. One straight forward approach to achieve this goal is to apply the previous methods for controlling smoke or water to clouds. However we found that this approach did not produce convincing results. The reason is that there are several physical processes, such as the phase transition from water vapor to water droplets (i.e. clouds), that are not present in other phenomena. We therefore developed a new method that controls the physical parameters affecting the cloud formation process. This results in natural cloud shapes and motion.

For the colors of clouds, we present a method for automatically adjusting the parameters such that the colors of the synthetic clouds are similar to a specified photograph of real clouds [3]. Our purpose is not to estimate physically correct parameters but to find the parameters that can produce an image that is visually similar to the clouds in the input photograph. We take into account important optical phenomena affecting the colors of clouds such as scattering and absorption of light inside the clouds. We use a color histogram to measure the visual difference between the synthetic image and the photograph. Solving the inverse rendering problem, however, is not trivial because the intensity of clouds is a highly nonlinear function of the parameters used to render them. Furthermore, there is seldom a unique solution to this problem: many different sets of parameters can produce similar images. We chose genetic algorithms (GAs) to address this problem because of their two capabilities: (1) they can find the optimal parameters efficiently even for such a highly nonlinear problem and (2) they can find a number of candidates for the optimal parameters during the optimization process.

In the following, we briefly explain these two methods and show some examples to demonstrate the effectiveness of our inverse approach.

2 Controlling Simulation of Cloud Formation

There are two requirements in designing our method for the inverse modeling of realistic cloud shapes: (1) realistic shapes of clouds have to be generated and (2) the shapes should closely match to the desired shape specified by the user. For the first

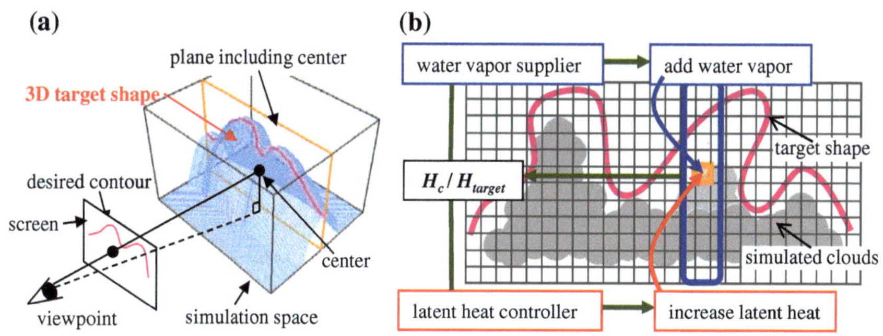

Fig. 1 Overview of our control method. **a** Generating 3D target shape from 2D input contour line. **b** Our control mechanism

requirement, we employ numerical simulation of cloud formation processes based on the atmospheric fluid dynamics. For the second requirement, we use a feedback control mechanism to automatically adjust some of the simulation parameters.

The physical processes for the cloud formation are as follows. First, where there are no clouds in the sky, the ground is heated by the sun. Then, the air near the ground is heated and air parcels start to move upward due to the thermal buoyance forces. The temperature of the rising air parcels decreases due to adiabatic cooling, so vapor in the air parcels causes a phase transition, coagulates, and water droplets are generated. The water droplets are perceived as the cloud. At that time when the phase transition occurs, the latent heat is liberated, which creates additional buoyancy forces and promotes further growth of the clouds. These processes can be expressed by five partial differential equations [5]. Two of them are the NS equations. The other three correspond to temperature, water vapor, and clouds. By simulating these processes numerically, realistic animation of the cloud formation can be created.

In order to control the simulation, the user specifies a contour line of the desired shape of clouds as indicated by the pink curve in Fig. 1a. Then a three-dimensional target shape is generated from the contour line. The simulation is controlled so that the difference between the target shape and the simulated clouds become zero. We developed two controllers, a latent heat controller and a water vapor supplier, to automatically adjusts the amount of the latent heat and the amount of water vapor to form the target shape. We chose the latent heat and the water vapor as control variables based on our experimental investigation. Other parameters remain fixed throughout the simulation. The control mechanism is shown in Fig. 1b. To measure the difference between the target shape and the simulated clouds, the height ratio of the height of the simulated clouds H_c to the height of the target shape H_{target} is calculated. The height ratio is fed back into the latent heat controller and the water vapor supplier. The latent heat controller increases the latent heat and the water vapor supplier adds water vapor in the regions where the clouds have not reached the top of the target shape. By combining these controllers, the vertical development of the clouds and the generation of clouds are controlled until the target shape is

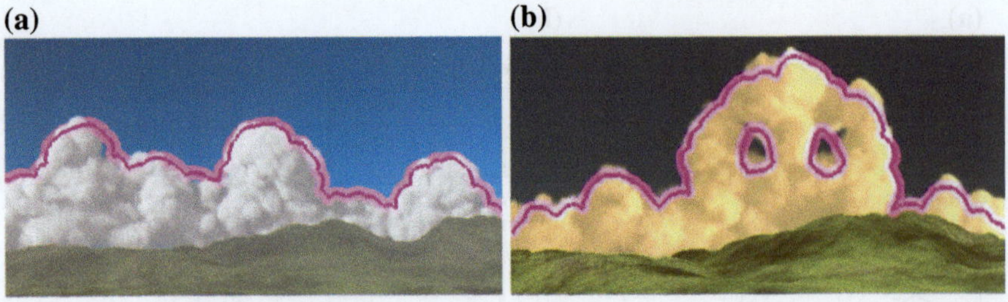

Fig. 2 Examples of clouds generated by our control method. **a** Typical cumulonimbus clouds. **b** Unnatural shape of clouds

formed. An important aspect of our control mechanism is that the simulation is *implicitly* controlled. No external forces nor cloud densities are explicitly generated. This prevents our controller from destructing the cloud dynamics and results in realistic cloud formation.

Figure 2 shows the clouds generated by our method. The shape of the clouds generated by our method is almost the same as the desired shape indicated by the pink curves. Figure 2a shows a typical shape of cumulonimbus clouds generated by using our control method. In Fig. 2b, the user specifies an unnatural shape of clouds, a skull. Our method can successfully generate realistic clouds even for this unnatural shape.

3 Automatic Adjustment of Parameters for Rendering Clouds

In order to display realistic clouds, the light scattering inside clouds need to be simulated. However, realistic images are not generated unless the user specifies good parameters used for the simulation of the light scattering. Our method addresses this problem by solving an inverse problem. That is, we let the computer search for the parameters that produce the realistic image. However, in order to do this, we have to tell the computer about the definition of the realism. In our approach, we use a photograph of real clouds and let the computer search for the parameters so that the appearance of the synthetic clouds look similar to those in the photograph.

An overview of our system is illustrated by Fig. 3. The inputs to our system are volume data representing the density distribution of synthetic clouds, and a photograph of real clouds. The direction of the sunlight and the camera parameters used to render synthetic clouds also need to be specified by the user. Our system then searches for the optimal parameters that minimize the following objective function O:

$$\arg\min_{\mathbf{c}} O(\mathbf{I}_{cg}(\mathbf{c}), \mathbf{I}_{usr}), \tag{1}$$

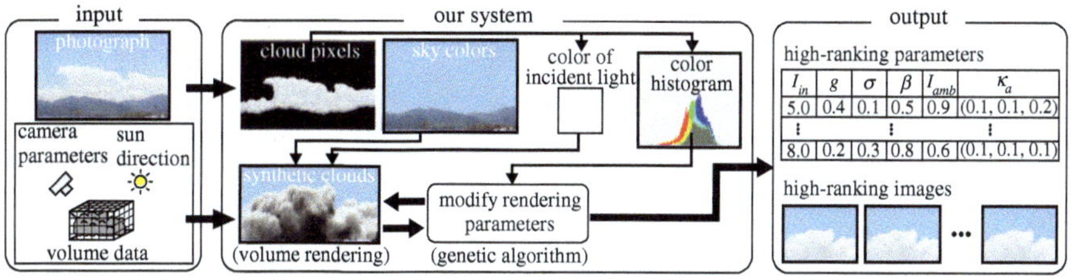

Fig. 3 Overview of our method for automatic adjustment of cloud colors

where **c** is a vector consisting of the parameters used for rendering the synthetic image $\mathbf{I}_{cg}$. $\mathbf{I}_{usr}$ is the photograph specified by the user. The objective function O measures the visual difference between $\mathbf{I}_{cg}$ and $\mathbf{I}_{usr}$. We use color histograms to compute the visual differences. O is defined by:

$$O = \frac{1}{3} \sum_{\lambda=R,G,B} \sum_{n=0}^{n_L-1} |h_{cg}(n,\lambda) - h_{usr}(n,\lambda)|, \tag{2}$$

where λ is the wavelength sampled at the wavelength corresponding to the RGB color channels, n_L is the number of intensity levels, and h_{cg} and h_{usr} represent histograms of $\mathbf{I}_{cg}$ and $\mathbf{I}_{usr}$, respectively. h_{cg} and h_{usr} are normalized by dividing them by the number of pixels. These histograms are computed using only the pixels corresponding to the clouds.

The intensity of clouds in the synthetic image $\mathbf{I}_{cg}$ is calculated based on the rendering equations for the clouds [2, 9, 11]. The intensity of clouds depends on many parameters, such as the intensities of the sunlight and the skylight, and the optical properties of atmospheric and cloud particles. In our method, the only light source illuminating the clouds is the sun. However, the intensity of light directly reaching the viewpoint from the sky behind the clouds is taken into account. The attenuation and scattering of light due to atmospheric particles between the clouds and the viewpoint are also taken into account.

The minimization problem defined above is solved by rendering the clouds repeatedly with various parameter settings using GAs. To render clouds, we take into account both single and multiple scattering. The scattering and absorption due to atmospheric particles between the clouds and the viewpoint are also taken into account. We employ the simplest model where the density of the atmospheric particles is assumed to be uniform and the intensity of scattered light due to atmospheric particles is assume to be constant. Under these assumptions, the intensity of light reaching the viewpoint for a pixel is a blended intensity of the clouds and the sky behind the clouds. Before using GAs, our system extracts cloud pixels from the input photograph and estimates the color of the incident light $c_{in}(\lambda)$ and the intensity of the sky behind the clouds $L_{sky}(q,\lambda)$. The color of the incident light is different from that of the sun because the sunlight is attenuated and scattered by atmospheric par-

Fig. 4 Examples of our method for adjusting parameters for rendering clouds. **a** Example of cumulonimbus clouds rendered with parameters determined by our method. **b** Application of our method to an animation of cumulonimbus clouds. **c** Composition of synthetic clouds onto a photograph

ticles before reaching the clouds. The color histogram of the input photograph is also calculated using the extracted cloud pixels. The rest of the parameters are then estimated using GAs. The images of the synthetic clouds are repeatedly created by using volume rendering techniques with different parameter settings. GAs compute the objective function for each of the candidate parameter sets to measure the quality of the parameters and modify the parameters. Each set of parameters is ranked by the objective function and high-ranking parameter sets are stored. The output of our system is a set of high-ranking parameters and their corresponding images.

Figure 4 shows examples of clouds rendered by using our method. Figure 4a shows an example of cumulonimbus clouds generated by fluid simulation [8]. The inset in each image is the input photograph of the clouds. By estimating the parameters for rendering the clouds, the subtle color variations observed in the photograph are reproduced in the synthetic clouds. Figure 4b shows an application of our method to create an animation of dynamic clouds. We used two parameters to render the daytime and the sunset clouds in Fig. 4a to create the animation with the position of the sun changing. The parameters were linearly interpolated. Figure 4b shows snapshots from the animation. Realistic color transitions are realized. Figure 4c shows an example of unnatural clouds. The clouds were generated using controlled simulation described in the previous section. In this example, we replaced the real clouds in the input photograph with the synthetic clouds, rendered using the optimized parameters. The synthetic clouds are naturally composited onto the real photograph.

4 Conclusion

We have presented an inverse design approach for visual simulation of clouds. Our approach takes two-dimensional information as input and computes three-dimensional information to create realistic images of clouds. For the cloud shapes, two-dimensional contour line of the desired shapes of clouds is used to generate three-dimensional shapes of clouds. For the cloud colors, a photograph of real clouds is used to find optimal parameters that can produce realistic images of the synthetic clouds. These inverse problems are highly nonlinear and hard to solve. We used the feedback control mechanism and genetic algorithms. The future work includes extension of the methods to visual simulation of other natural phenomena such as fire, smoke, water, and so on.

References

1. Bouthors A, Neyret F, Max N, Bruneton E, Crassin C (2008) Interactive multiple anisotropic scattering in clouds. In: Proceedings of ACM symposium on interactive 3D graphics and games, pp 173–182
2. Cerezo E, Perez F, Pueyo X, Seron FJ, Sillion FX (2005) A survey on participating media rendering techniques. Vis Comput 21(5):303–328
3. Dobashi Y, Iwasaki W, Ono A, Yamamoto T, Yue Y, Nishita T (2012) An inverse problem approach for automatically adjusting the parameters for rendering clouds using photographs. ACM Trans Graph 31(6):Article 145
4. Dobashi Y, Kaneda K, Yamashita H, Okita T, Nishita T (2000) A simple, efficient method for realistic animation of clouds. In: Proceedings of ACM SIGGRAPH 2000, pp 19–28
5. Dobashi Y, Kusumoto K, Nishita T, Yamamoto T (2008) Feedback control of cumuliform cloud formation based on computational fluid dynamics. ACM Trans Graph 27(3):Article 94
6. Gardner GY (1985) Visual simulation of clouds. Comput Graph (Proc SIGGRAPH 1985) 19(3):297–304
7. Harris MJ, Lastra A (2001) Real-time cloud rendering. Comput Graph Forum 20(3):76–84
8. Miyazaki R, Dobashi Y, Nishita T (2002) Simulation of cumuliform clouds based on computational fluid dynamics. In: Proceedings of EUROGRAPHICS 2002 short presentations, pp 405–410, Aug 2002
9. Nishita T, Dobashi Y, Nakamae E (1996) Display of clouds taking into account multiple anisotropic scattering and sky light. In: Proceedings of ACM SIGGRAPH 1996, pp 379–386
10. Yue Y, Iwasaki K, Chen B-Y, Dobashi Y, Nishita T (2010) Unbiased, adaptive stochastic sampling for rendering inhomogeneous participating media. ACM Trans Graph 29(6):Article 177
11. Zhou K, Ren Z, Lin S, Bao H, Guo B, Shum H-Y (2008) Real-time smoke rendering using compensated ray marching. ACM Trans Graphics 27(3):Article 36

Generating Flow Fields Variations Using Laplacian Eigenfunctions

Syuhei Sato, Yoshinori Dobashi, Kei Iwasaki, Hiroyuki Ochiai and Tsuyoshi Yamamoto

Abstract The visual simulation of fluids has become an important element in many applications, such as movies and computer games. In these applications, large-scale fluid scenes, such as fire in a village, are often simulated by repeatedly rendering multiple small-scale fluid flows. In these cases, animators are requested to generate many variations of a small-scale fluid flow. This chapter presents a method to help animators meet such requirements. Our method enables the user to generate flow field variations from a single simulated dataset obtained by fluid simulation. The variations are generated in both the frequency and spatial domains. Fluid velocity fields are represented using Laplacian eigenfunctions which ensure that the flow field is always incompressible. Using our method, the user can easily create various animations from a single dataset calculated by fluid simulation.

Keywords Flow field · Variation synthesis · Laplacian eigenfunctions · Amplitude modulation · Resizing simulation space

S. Sato (✉) · T. Yamamoto
Graduate School of Information Science and Technology, Hokkaido University,
Kita 14 Nishi 9, Kita-ku, Sapporo 060-0814, Japan
e-mail: sato@ime.ist.hokudai.ac.jp

T. Yamamoto
e-mail: yamamoto@ime.ist.hokudai.ac.jp

Y. Dobashi
Graduate School of Information Science and Technology, Hokkaido University/JST CREST,
Kita 14 Nishi 9, Kita-ku, Sapporo 060-0814, Japan
e-mail: doba@ime.ist.hokudai.ac.jp

K. Iwasaki
Wakayama University, 930, Sakaedani, Wakayama 640-8510, Japan
e-mail: iwasaki@sys.wakayama-u.ac.jp

H. Ochiai
Institute of Mathematics for Industry, Kyushu University/JST CREST, 744, Motooka,
Nishi-ku, Fukuoka 819-0395, Japan
e-mail: ochiai@imi.kyushu-u.ac.jp

K. Anjyo (ed.), *Mathematical Progress in Expressive Image Synthesis I*,
Mathematics for Industry 4, DOI: 10.1007/978-4-431-55007-5_13,

1 Introduction

The visual simulation of fluids has become one of the most important research topics in computer graphics. Many methods have been proposed for simulating smoke, water, fire, etc [2]. Most of the recent methods are based on computational fluid dynamics to create realistic animations and these are used in many applications such as movies and computer games. However, one of the problems is the expensive computational cost. In those entertainment applications, similar fluid animations are often required. Such examples include multiple explosions caused by missile attack, many flowing rivers, multiple houses on fire, and smoke rising from multiple chimneys. Using the same fluid animation repeatedly degrades the realism of a synthetic scene. Therefore, animators have to create multiple fluid animations with different motions. This is achieved by repeating the fluid simulation many times with different parameter settings. However, adjusting the simulation parameters to create such similar animations is very difficult and huge computational costs are required. Our research goal is to address this problem.

Procedural methods can generate similar multiple flows with relatively low computational cost. For example, curve-based methods for creating various fire animations have been proposed [3, 4]. These methods generate user-designed fire animations by deforming a curve representing the route of the fire. Some researchers have developed procedural methods for modeling flow using turbulent noise functions [1, 5]. With these methods, animators can easily generate turbulent motion and multiple animations with similar motion can be created at low cost. However, these procedural methods do not use fluid simulation to generate the fluid flow, making the animations less realistic than those created using physically-based simulation.

To address this problem, we propose a method for generating various different fluid animations from a single simulated dataset.Our method is based on grid-based fluid simulation and is suitable for generating animations of divergence-free flow such as smoke and fire. The key concept behind our method is to represent the input velocity fields using divergence-free basis functions. We use Laplacian eigenfunctions [7] as the divergence-free basis functions. this basis functions enforce divergence-free condition to flow fields, and flow fields are decomposed into frequency components. The variations are generated by randomly modulating the coefficient of frequency components. Furthermore, our system resizes the simulation space to generate variations in the spatial domain. The flow field in the resized space is calculated by solving a minimization problem. By combining these two methods, various fluid flows can be generated from a single simulated flow field.

2 The Method

Figure 1 shows an overview of our method. Two types of processes are used to generate the variations. The first one is amplitude modulation which generates variations of the flow field $\mathbf{v}_{mod}(t_n)$ in the frequency domain. This process modulates

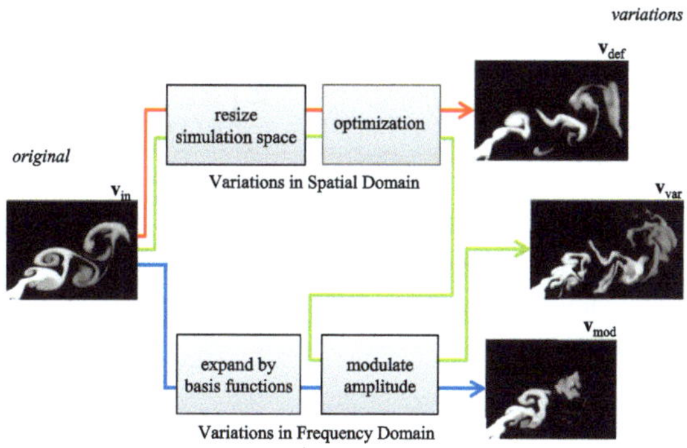

Fig. 1 Overview of our method

the coefficients calculated by expanding the input flow field $\mathbf{v}_{in}(t_n)$ into divergence-free basis functions. The other one is resizing the simulation space which generates variations of the flow field $\mathbf{v}_{def}(t_n)$ in the spatial domain. In this process, the size of the simulation space is changed randomly. Our system then solves a minimization problem subject to the resized velocity fields, and generates the resultant flow fields. These processes can be combined and a flow field $\mathbf{v}_{var}(t_n)$ is generated. In the following subsections, we describe the details of the above two processes.

2.1 Variations in the Frequency Domain

First, we expand the input flow field $\mathbf{v}_{in}(t_n)$ into divergence-free basis functions $\mathbf{b}_i$. Our system calculates coefficients $w_i(t_n)$ for the i-th basis function $\mathbf{b}_i$ using the following equation.

$$w_i(t_n) = \int_{\mathbf{x}\in\Omega} \mathbf{v}_{in}(t_n) \cdot \mathbf{b}_i d\mathbf{x}, \tag{1}$$

where Ω is the entire domain of the input flow field and $\cdot$ is the dot product between two vectors, and t_n are discrete time steps ($n = 0, 1, \cdots$). Using the coefficients $w_i(t_n)$, we represent the input flow field $\mathbf{v}_{in}(t_n)$ by a linear combination of basis functions $\mathbf{b}_i$, as follows,

$$\mathbf{v}_{in}(t_n) = \sum_{i=0}^{N-1} w_i(t_n)\mathbf{b}_i, \tag{2}$$

where N is the number of basis functions. We use Laplacian eigenfunctions for the basis functions, $\mathbf{b}_i$ [7]. This enforces the divergence-free condition on the flow field. In addition, flow fields can be decomposed into frequency components by using this basis functions. For a 2D flow field, the Laplacian eigenfunctions are defined on a 2D rectangular grid. The 2D Laplacian eigenfunctions $\mathbf{b}_k = (f_{k_1,k_2}(x, y), g_{k_1,k_2}(x, y))$ in a rectangular region $[0, 0] \times [s_x, s_y]$ are defined by the following equations.

$$f_{k_1,k_2}(x,y) = \frac{1}{\lambda} k_2 \sin(k_1 \pi x / s_x) \cos(k_2 \pi y / s_y), \tag{3}$$

$$g_{k_1,k_2}(x,y) = \frac{-1}{\lambda} k_1 \cos(k_1 \pi x / s_x) \sin(k_2 \pi y / s_y), \tag{4}$$

where k_1, k_2 are wave numbers, and λ is an eigenvalue defined by $\lambda = k_1^2 + k_2^2$.

By modulating $w_i(t_n)$, we generate variations of the flow fields $\mathbf{v}_{mod}(t_n)$ in the frequency domain. Our method modulates $w_i(t_n)$ for each frequency component as follows,

$$\mathbf{v}_{mod}(t_n) = \sum_{i=0}^{N-1} g_i w_i(t_n) \mathbf{b}_i, \tag{5}$$

where g_i represents the gain with which $w_i(t_n)$ is modulated.In our method, g_i is generated randomly.

2.2 Variations in the Spatial Domain

In this subsection, we describe a method for generating flow field variations in the spatial domain. Our method changes the size of the simulation space by arbitrarily expanding or contracting it. Figure 2 shows the method used to resize the simulation space. First, a direction and a line that cuts through the input velocity field are determined by random numbers. In Fig. 2, the cut direction is shown as being horizontal. Next, a distance to expand or contract the simulation space is determined, again using a random number. In the expansion process, the simulation space is divided by the cut line into two domains. These domains are separated and new grid points are inserted between them (Fig. 2). In the case of contraction, two cut lines are specified, and then the grid points between the two cut lines are removed as shown in Fig. 2. After resizing the simulation space, we prepare a new grid for the resized simulation space. The number of grid points for the new grid is N_{def}. The input velocities are copied to the new grid according to the resizing information of the simulation space. The resized velocity field is denoted by $\mathbf{v'}_{in}$. Then the coefficients $\mathbf{w}'(t_n) = (w'_0(t_n), w'_1(t_n), \ldots, w'_{N-1}(t_n))$ are calculated by solving the following minimization problem.

$$\arg\min_{\mathbf{w}'(t_n)} (E(t_n) + \alpha \sum_{i=0}^{N-1} w_i'^2(t_n)), \tag{6}$$

$$E(t_n) = \sum_{\mathbf{x} \in \Omega_{in}} |\mathbf{v'}_{in}(\mathbf{x}, t_n) - \sum_{i=0}^{N-1} w'_i(t_n) \mathbf{b'}_i(\mathbf{x})|^2,$$

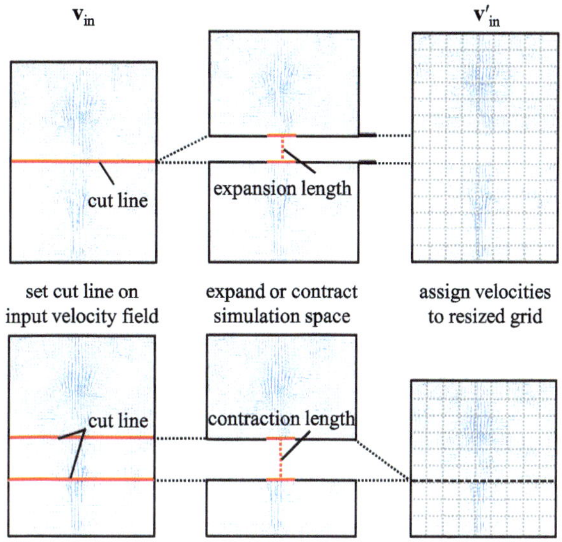

Fig. 2 Expanding and contracting an input velocity field

where Ω_{in} is the domain where the input velocities $\mathbf{v}_{in}$ are assigned on $\mathbf{v}'_{in}$, and α is a user-specified constant used to adjust the influence of the second term which is called the regularization term. $\mathbf{b}'_i$ is defined on the grid of the resized simulation space. $E(t_n)$ measures the difference between the resized velocities and the resulting flow field. By solving the above equation, variations of the flow fields $\mathbf{v}_{def}(t_n)$ in the spatial domain are synthesized.

$$\mathbf{v}_{def}(t_n) = \sum_{i=0}^{N-1} w'_i(t_n)\mathbf{b}'_i. \tag{7}$$

Given the definition of the error function $E(t_n)$ in the previous paragraph, we can now solve the minimization problem given in Eq. (6). By taking the derivative of Eq. (6) with respect to the coefficient $w'_i(t_n)$, we obtain the following matrix equation.

$$(\mathbf{A} + \alpha\mathbf{I})\mathbf{w}'(t_n) = \mathbf{c}(t_n), \tag{8}$$

where $\mathbf{A}$ and $\mathbf{I}$ are $N \times N$ matrices, and $\mathbf{w}'(t_n)$ and $\mathbf{c}(t_n)$ are N dimensional column vectors. $\mathbf{I}$ is the identity matrix. $\mathbf{A}$ and $\mathbf{c}$ are related to E. The (i, j)-th element a_{ij} of $\mathbf{A}$ and the i-th element c_i of $\mathbf{c}(t_n)$ are given respectively by:

$$a_{ij} = \sum_{\mathbf{x}\in\Omega_{in}} \mathbf{b}'_i(\mathbf{x}) \cdot \mathbf{b}'_j(\mathbf{x}),$$

$$c_i = \sum_{\mathbf{x}\in\Omega_{in}} \mathbf{v}'_{in}(\mathbf{x}, t_n) \cdot \mathbf{b}'_i(\mathbf{x}),$$

Table 1 Parameter settings and computation times

	N	N_{def}	T		N	N_{def}	T
Figure 1	1024	–	120	Figure 5b, c	1024	256×320	140
Figure 4b	400	64×96	10	Figure 5e, f	1024	256×192	80

N is the number of basis functions, N_{def} is the number of grid points for resized velocity fields. T is the time for updating the flow field, measured in milliseconds

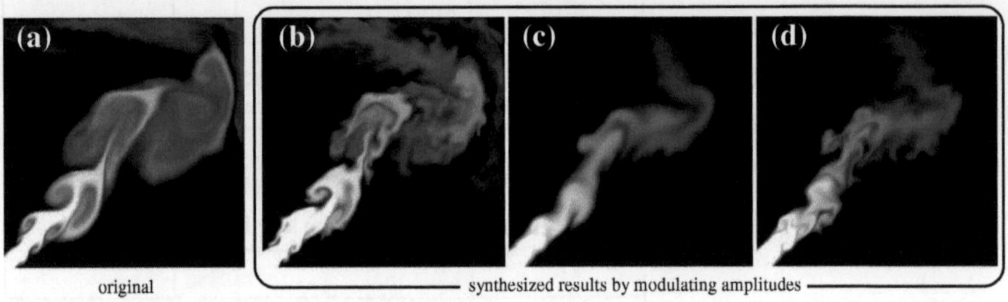

Fig. 3 Modulating amplitudes of a smoke animation by our method

$\mathbf{A}$ is a full matrix as shown by the above equation. In order to compute the dot product between the functions in the above equations, $\mathbf{v}'_{in}$ and $\mathbf{b}'_i$ are sampled on the grid of the resized simulation space. Then, an approximation of the dot product is computed using the sampled values.

The coefficient vector $\mathbf{w}'(t_n)$ is then obtained by solving the matrix equation, Eq. (8). To solve the equation efficiently, we use the LU decomposition technique [6]. The LU decomposition of the matrix $(\mathbf{A}+\alpha\mathbf{I})$ is firstly computed and the weight vector $\mathbf{w}'(t_n)$ is efficiently obtained using the decomposed matrix.

3 Results

This section shows some examples created using our method. We used a desktop PC with an Intel Core i7 2600K CPU, 16GB memory, and an NVIDIA GeForce GTX 680 GPU to compute all the examples shown in this section. The parameter settings and timing information are summarized in Table 1. For examples of Figs. 3 and 5, the number of grid points for the input velocity fields was 256×256. In Fig. 4, the input velocity field was computed on a 64×64 grid.

Figure 3 shows examples of the amplitude modulation. Note that the flow field is visualized by advecting the smoke density. The source of the smoke is located at the bottom-left corner of the simulation space. Figure 3a shows the input smoke animations. Figure 3b–d show the variations created by modulating the amplitudes of the input animations. (b) is created by increasing the high-frequency components of the input velocity fields. In (c), the low-frequency components are reduced. (d) is

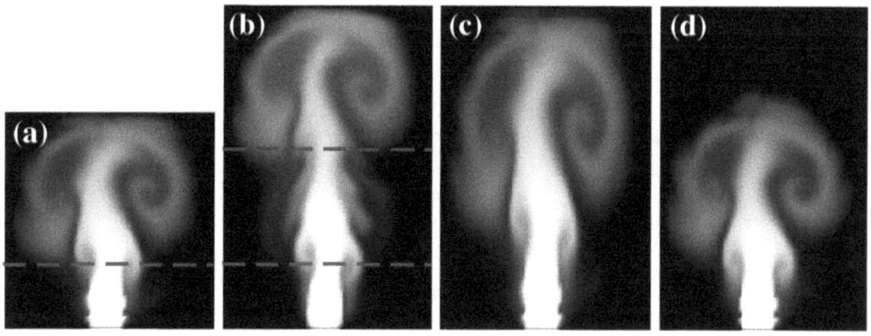

Fig. 4 Comparison of results using different methods

created by applying the modulations of both (b) and (c). As shown in these examples, our method can generate variations in the frequency domain.

Figure 4 shows a comparison of results generated by using other resizing approaches with a result using our method. Figure 4a shows the original flow field computed on a 64×64 grid. A cut line is shown by a red dotted line in Fig. 4a. The result by using our method is shown in Fig. 4b. Our method successfully created a continuous flow field and the flow is similar to the original one. Figure 4c shows a flow field created by warping the original flow field into the resized simulation space. In this case, the flow is deformed from the original flow field. Figure 4d shows a result obtained by re-simulating the flow field in the resized simulation space with the same parameter settings as in Fig. 4a. In this case, the smoke does not rise into the higher region unless we adjust the simulation parameters adequately through a trial-and-error process.

Figure 5 shows the results obtained using our method. Figure 5a, d show the original flow field computed by the fluid simulation. The cut lines in Fig. 5a, d are shown by red dotted lines. In Fig. 5b, c, the simulation space has been expanded. In Fig. 5e, f, the simulation space has been contracted. The amount of expansion and contraction is set to a quarter of the vertical height of the input animation. Furthermore, Fig. 5c, f are created by additionally applying amplitude modulation so that the high-frequency components are enhanced. Using our method continuous flow fields can be successfully created. In addition, we can generate variations in both the frequency and spatial domains.

4 Conclusion

We have proposed a method for synthesizing variations of flow fields. Our method can generate variations in both the frequency and spatial domains. The flow fields are represented by Laplacian eigenfunctions, and the variations in the frequency domain are generated by modulating the coefficients of the basis functions. In addition, we can generate variations in the spatial domain by solving a minimization problem

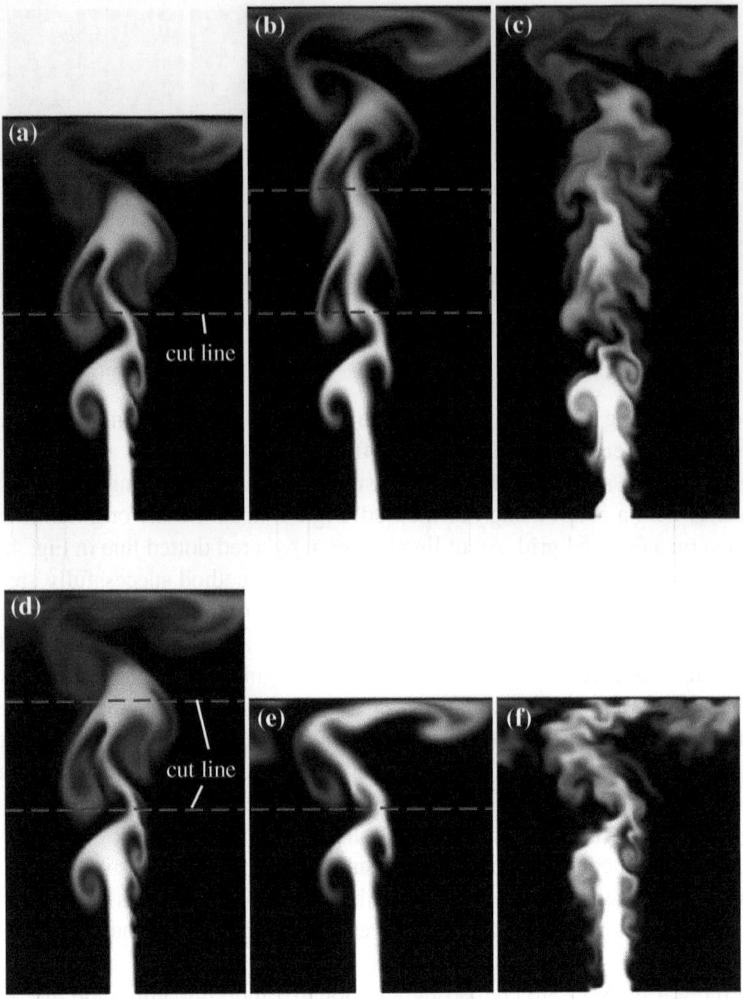

Fig. 5 Resizing simulation space of input smoke animation

after resizing the velocity field. We demonstrated the capabilities of our method with a set of examples. In future work, we are planning to extend our method to 3D flow fields.

One of the limitations of our method is that the computational cost is proportional to the number of basis functions. The level of detail in the flow generated by our method depends on the number of basis functions used. If the number of basis functions is large, high computational costs are required to calculate $\mathbf{A}$, $\mathbf{c}(t_n)$ and to reconstruct the flow fields using Eqs. (5) and (7). The user can generate variations of detailed flow fields at the expense of increased computational and storage costs.

Another limitation is the fact that the flow fields generated by our method might not conform to the laws of fluid flow, if the degree of modulation by g_i is too large. To address this problem, we calculate the Navier–Stokes equations for velocity fields generated by our method, then compare two velocity fields. By this comparison, we can evaluate whether or not the flow generated by our method follows the laws of fluid flow. In future work, we propose to evaluate these flow fields.

References

1. Bridson R, Hourihan J, Nordenstam M (2007) Curl-noise for procedural fluid flow. ACM Trans Graph 26(3):Article 46
2. Bridson R (2008) Fluid simulation for computer graphics. AK Peters
3. Fuller AR, Krishnan H, Mahrous K, Hamann B, Joy KI (2007) Real-time procedural volumetric fire. In: Proceeding of the 2007 symposium on Interactive 3D graphics and games, pp 175–180
4. Lamorlette A, Foster N (2002) Structural modeling of flames for a production environment. ACM Trans Graph 21(3):729–735
5. Patel M, Taylor N (2005) Simple divergence-free fields for artistic simulation. J Graph GPU Game Tools 10(4):49–60
6. Press WH, Teukolsky SA, Vetterling WT, Flannery BP (2007) Numerical recipes, 3rd edn: The art of scientific computing. Cambridge University Press, Cambridge
7. Witt TD, Lessig C, Fiume E (2012) Fluid simulation using Laplacian eigenfunctions. ACM Trans Graph 31(1):Article 10

Blood Flow Analysis Using Medical Imaging Data and Streamline Visualization

Hiroshi Suito and Takuya Ueda

Abstract A numerical simulation of blood flow in the thoracic aorta is presented. Patient-specific aorta shapes are used in a centerline-fitted curvilinear coordinate system in which the Navier–Stokes equation is discretized using finite-difference approximation with immersed boundary method. Numerical results are visualized by drawing instantaneous streamlines to exhibit the flow structures. Dean vortices and the transition from them to a swirling flow are observed by comparing the original shapes and projected shapes that have no torsion.

Keywords Blood flow · Aortic aneurysm · Numerical simulation · Scientific visualization · Swirling flow

1 Introduction

This chapter presents an investigation of blood flows in the thoracic aorta which is related to thoracic aortic aneurysms [2, 6].

In recent years, computational fluid dynamics for blood flow has been of wide interest. From the viewpoint of computational methods, two main discretization strategies exist: unstructured and structured meshes. When attempting to compute flows using structured meshes, the most important issue is the manner of representing the geometry in the structured mesh system. The strategies to resolve this problem are

---

H. Suito (✉)
Graduate School of Environmental and Life Sciences, Okayama University/JST CREST, 3-1-1 Tsushima-naka, Okayama 700-8530, Japan
e-mail: suito@okayama-u.ac.jp

T. Ueda
Department of Radiology, St. Luke's International Hospital/JST CREST, 9-1 Akashi-cho, Tokyo 104-8560, Japan
e-mail: takueda-rad@umin.ac.jp

K. Anjyo (ed.), *Mathematical Progress in Expressive Image Synthesis I*, Mathematics for Industry 4, DOI: 10.1007/978-4-431-55007-5_14,

classifiable into two categories: body-fitted and immersed-boundary approaches [3]. An advantage of body-fitted mesh is its high accuracy, although mesh generation near a complex-shaped boundary is often difficult. In contrast, in the immersed boundary method, mesh generation is simple because a uniform orthogonal mesh is useful and because the wall geometry is represented by a distribution of a characteristic function. This study uses a hybrid approach that is intermediate of the body-fitted and immersed boundary method.

As described in this chapter, we present some numerical simulations based on the hybrid method described above. In addition, we show streamline visualization to elucidate the flow structures.

2 Representation of Aorta Morphology

To extract the (x, y, z)—coordinate of the centerline and radius at position s from CT images, median axis transform technique [4, 5] is applied, where s is a length along the centerline from the proximal end of the aorta. Branches are neglected in this study. Then, a curvilinear coordinate system (ξ, η, ζ) is generated, for which the ζ axis is nearly parallel to the aorta centerline. Therefore, (ξ, η, ζ) is a centerline-fitted coordinate system that represents the coordinate transformation from computational space (ξ, η, ζ) to physical space (x, y, z). The ζ−axis is set not to be strictly parallel but to be nearly parallel to the centerline, which means that the original centerline geometry is smoothed slightly by the Gaussian filter. This treatment avoids numerical instabilities arising from the severe skewness of finite difference meshes.

On each mesh point defined in the preceding step, the characteristic function is generated, which takes the value of 0 inside of the aorta and value 1 outside of the aorta. It varies gradually near the vessel wall with certain smoothness. Figure 1 portrays a centerline and radius data at several points (left), a finite difference mesh with the centerline and radius data (center), and a contour surface of the characteristic function with the finite difference mesh (right).

3 Governing Equations

As governing equations, the incompressible viscous Navier–Stokes equations and the continuity equation

$$\begin{cases} \frac{\partial u}{\partial t} + (u \cdot \nabla)\, u = -\frac{1}{\rho}\nabla p + \nu\nabla^2 u - \frac{1}{\varepsilon_f}\lambda u \\ \nabla \cdot u = 0 \end{cases}, \tag{1}$$

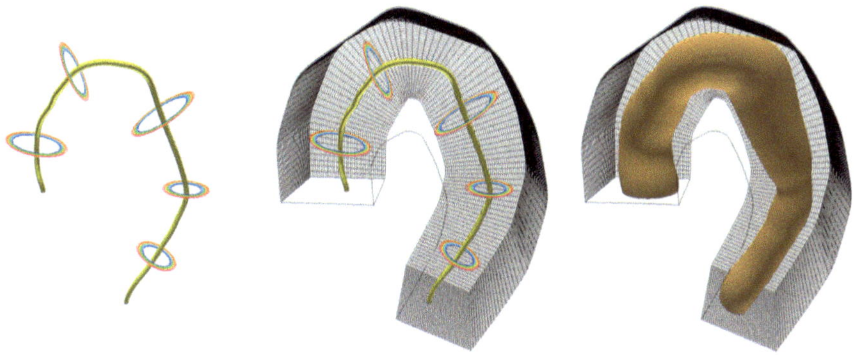

Fig. 1 Surface shape reconstruction procedure

are used with initial and boundary conditions where u, p, t, ρ, and ν respectively denote velocity, pressure, time, density, and kinematic viscosity. The last term of the Navier–Stokes equations is a drag force term that is proportional to the fluid velocity because of the immersed boundary method of the discrete forcing type [3]. This force acts only out of the aorta. Therefore, the flow velocity out of the aorta becomes zero and the aorta wall shape is represented.

4 Visualization of Numerical Results

To elucidate the flow field in the thoracic aorta, computed flow fields are visualized by drawing instantaneous streamlines at $t = t_1$, which is defined as

$$\frac{dx}{u(x, y, z, t_1)} = \frac{dy}{v(x, y, z, t_1)} = \frac{dz}{w(x, y, z, t_1)}, \tag{2}$$

where u, v, w, and dx, dy, dz respectively represent the x, y, z components of the fluid velocity and the line segment. The instantaneous streamlines differ from streak lines or particle paths, which show different views in unsteady flows.

Figure 2 shows instantaneous streamlines immediately after the peak systolic phase of the heart beat where the widths and colors of the streamlines represent the velocity magnitude. In Fig. 2, several characteristic flows are visible. For example, swirling flows are apparent in the descending aorta. Figure 3 shows the instantaneous streamlines for another case, which show similar characteristics to those of the previous case.

Then we examine the effect of torsion of the aorta. The original shape is projected onto an averaged plane of curvature so that the projected shape has no torsion, i.e., the centerline is on a flat plane.

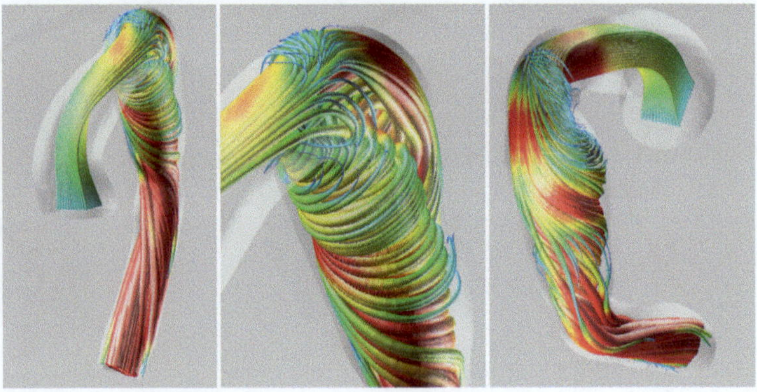

Fig. 2 Instantaneous streamlines (Case A)

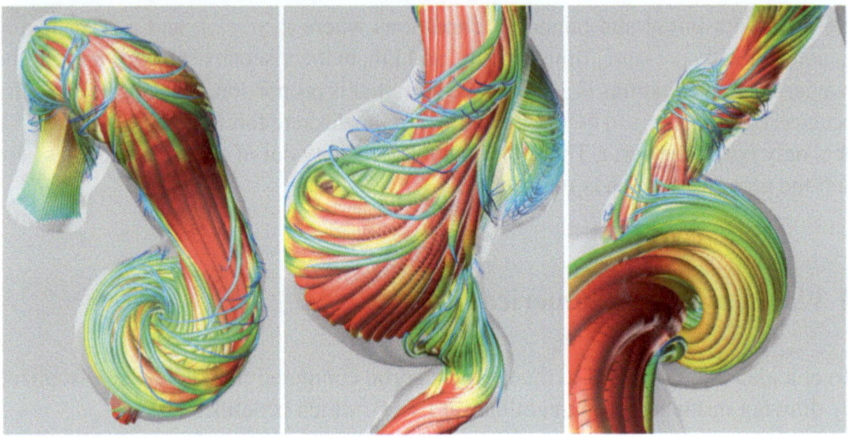

Fig. 3 Instantaneous streamlines (Case B)

Figures 4 and 5 show instantaneous streamlines for Case C, with side and back views of the original and projected shapes. Generally speaking, a set of twin symmetric vortices are formed in various curved tubes, which are induced by a centrifugal force depending on the axial flow distribution. These twin vortices are called Dean vortices, which play an important role in curved tubes. This kind of curvature effect is characterized by the Dean number [1]. In the projected shape (Fig. 4), the symmetric Dean vortices are apparent, although they are deformed and merge to one swirling flow in the original shape with torsion (Fig. 5). This phenomenon is apparently important because the wall shear stress distribution is changed dramatically by such a flow structure change. Furthermore, wall shear stress exerted on the aortic wall has a strong relation to the formation and development of aortic aneurysms.

side view back view

Fig. 4 Instantaneous streamlines (Case C: projected shape)

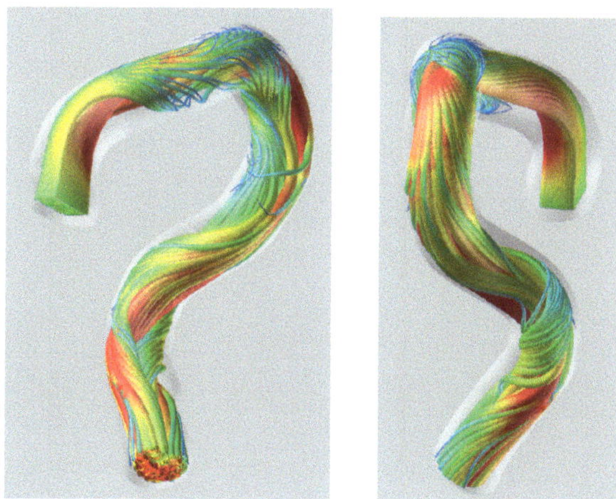

side view back view

Fig. 5 Instantaneous streamlines (Case C: original shape)

5 Conclusions

In this chapter, we presented streamline visualization to elucidate the flow structures. Instantaneous streamlines are suitable for elucidating and depicting the vortical flow structures in curved tubes, such as Dean vortices and swirling flows. Nevertheless, more sophisticated visualization techniques are highly desirable because the flows in three-dimensional curved vessels are extremely complicated.

Acknowledgments This work was supported by the JST-CREST Mathematics program.

References

1. Berger SA, Talbot L, Yao L-S (1983) Flow in curved pipes. Ann Rev Fluid Mech 15:461–512
2. Isselbacher EM (2005) Thoracic and abdominal aortic aneurysms. Circulation 111:816–828
3. Mittal R, Iaccarino G (2005) Immersed boundary methods. Ann Rev Fluid Mech 37:239–261
4. Rubin GD, Paik DS, Johnson PC, Napel S (1998) Measurement of the aorta and its branches with helical CT. Radiology 206:823–829
5. Tillich M, Bell RE, Paik DS, Fleischmann D, Sofilos MC, Logan LJ, Rubin GD (2001) Iliac arterial injuries after endovascular repair of abdominal aortic aneurysms: correlation with iliac curvature and diameter. Radiology 219:129–136
6. Ueda T, Fleischmann D, Rubin GD, Dake MD, Sze DY (2008) Imaging of the thoracic aorta before and after stent-graft repair of aneurysms and dissections. Semin Thorac Cardiovasc Surg 20(4):348–357

Part IV
Deformation and Locomotion

Discrete Isoperimetric Deformation of Discrete Curves

Jun-ichi Inoguchi, Kenji Kajiwara, Nozomu Matsuura and Yasuhiro Ohta

Abstract We consider isoperimetric deformations of discrete plane/space curves. We first give a brief review of the theory of isoperimetric deformation of smooth curves, which naturally gives rise to the modified KdV (mKdV) equation as a deformation equation of the curvature. We then present its discrete model described by the discrete mKdV equation, which is formulated as the isoperimetric equidistant deformation of discrete curves. We next give a review of isoperimetric and torsion-preserving deformation of smooth space curves with constant torsion which is described by the mKdV equation. We formulate a discrete analogue of it as the isoperimetric, torsion-preserving and equidistant deformation on the osculating planes of space discrete curves. The deformation admits two discrete flows, namely by the discrete mKdV equation and by the discrete sine-Gordon equation. We also show that one can make an arbitrary choice of two flows at each step, which is controlled by tuning the deformation parameters appropriately.

J. Inoguchi
Department of Mathematical Sciences, Yamagata University, 1-4-12, Kojirakawa, Yamagata 990-8560, Japan
e-mail: inoguchi@sci.kj.yamagata-u.ac.jp

K. Kajiwara (✉)
Institute of Mathematics for Industry, 744, Motooka, Nishi-ku, Fukuoka 819-0395, Japan
e-mail: kaji@imi.kyushu-u.ac.jp

N. Matsuura
Department of Applied Mathematics, Fukuoka University, 8-19-1, Nanakuma, Jonan-ku, Fukuoka 814-0180, Japan
e-mail: nozomu@fukuoka-u.ac.jp

Y. Ohta
Department of Mathematics, Kobe University, 1-1, Rokkodai, Nada-ku, Kobe 657-8501, Japan
e-mail: ohta@math.sci.kobe-u.ac.jp

K. Anjyo (ed.), *Mathematical Progress in Expressive Image Synthesis I*,
Mathematics for Industry 4, DOI: 10.1007/978-4-431-55007-5_15,

Keywords Curve motion · Discrete curve · Integrable systems · Discrete differential geometry · Modified KdV equation · Discrete modified KdV equation · Discrete sine-Gordon equation

1 Introduction

It is well-known that there are deep connections between the differential geometry and the theory of integrable systems. The theory of deformation of plane/space curves is a typical example, where the Frenet frame of curves and its deformation are described by system of linear partial differential equations. The modified KdV (mKdV) or the nonlinear Schrödinger (NLS) equation and their hierarchies arise naturally as the compatibility condition, as shown by various researchers including Hasimoto and Lamb [2, 6, 7, 18, 19, 22]. As the development of the discrete differential geometry, various continuous deformations of discrete curves have been studied, for example, in [3, 10, 11, 13, 14, 16, 21, 23]. The discrete deformations of discrete curves have not been studied well compared to the continuous deformations; deformation of discrete curves on sphere by the discrete sine-Gordon equation [4], and deformation of space discrete curves by the discrete NLS equation [12, 24] are such examples.

Recently we have presented the discrete deformation of plane discrete curves by the discrete mKdV equation [15, 20], and that of space discrete curves with constant torsion by the discrete mKdV and the discrete sine-Gordon equations. In this paper, we give a brief review of those discrete deformations of plane/space discrete curves.

2 Plane Curves

2.1 *Isoperimetric Deformation of Smooth Plane Curves*

It is well-known that the smooth plane/space curves admit the isoperimetric motion described by the *modified Korteweg-de Vries (mKdV) equation* [6, 18]

$$\dot{\kappa} = \frac{3}{2}\kappa^2\kappa' + \kappa''', \tag{1}$$

which is one of the most typical integrable systems. Here $\kappa = \kappa(x,t)$, $\dot{}$ and $'$ denote t- and x-derivatives, respectively. Let $\gamma(x,t) \in \mathbb{R}^2$ be an arc-length parameterized plane curve, x be the arc-length so that $|\gamma'| = 1$, and t be a deformation parameter. Let $T(x,t)$ and $N(x,t)$ be the *tangent vector* and the *normal vector* defined by

$$T(x,t) = \gamma'(x,t),\ N(x,t) = R\left(\frac{\pi}{2}\right) T(x,t), \tag{2}$$

respectively, where

$$R(\theta) = \begin{bmatrix} \cos\theta & -\sin\theta \\ \sin\theta & \cos\theta \end{bmatrix}, \tag{3}$$

and we define the *Frenet frame* $\Phi(x,t) \in \mathrm{SO}(2)$ by $\Phi(x,t) = [T(x,t), N(x,t)]$. Then Φ satisfies the *Frenet formula*

$$\Phi'(x,t) = \Phi(x,t)L(x,t), \quad L = \begin{bmatrix} 0 & -\kappa \\ \kappa & 0 \end{bmatrix}, \tag{4}$$

where $\kappa = \kappa(x,t)$ is the *curvature*. We consider the following deformation of the curve given by

$$\dot{\gamma} = \frac{\kappa^2}{2}T + \kappa' N, \tag{5}$$

which can be expressed in terms of the Frenet frame as

$$\dot{\Phi(x,t)} = \Phi(x,t)M(x,t), \quad M = \begin{bmatrix} 0 & -\kappa'' - \frac{\kappa^3}{2} \\ \kappa'' + \frac{\kappa^3}{2} & 0 \end{bmatrix}. \tag{6}$$

One can show by direct calculation that the deformation defined by (5) is *isoperimetric*, namely, $|\gamma'(x,t)| = 1$ for all t. The compatibility condition $\dot{L} - M' = [L, M]$ yields the mKdV equation (1). The linear system (4) and (6) is referred to as the *auxiliary linear problem* or the *Lax pair* of the mKdV equation (1).

2.2 Discrete Isoperimetric Deformation of Discrete Plane Curves

In this section, we discuss the discrete isoperimetric deformation of plane discrete curves by the discrete mKdV equation according to [15, 20]. For $\gamma_n \in \mathbb{R}^2$ ($n \in \mathbb{Z}$), if any three consecutive points $\gamma_{n-1}, \gamma_n, \gamma_{n+1}$ are not colinear, we call γ_n a *discrete plane curve*. Let $\gamma_n^m \in \mathbb{R}^2$ be a discrete plane curve, where $m \in \mathbb{Z}$ is the discrete deformation parameter. We define the *tangent vector* T_n^m and the *normal vector* N_n^m by

$$T_n^m = \frac{\gamma_{n+1}^m - \gamma_n^m}{\left|\gamma_{n+1}^m - \gamma_n^m\right|}, \quad N_n^m = R\left(\frac{\pi}{2}\right) T_n^m, \tag{7}$$

respectively. Introducing the *discrete Frenet frame* $\Phi_n^m \in \mathrm{SO}(2)$ by $\Phi_n^m = \left[T_n^m, N_n^m\right]$, it follows from the definition *the discrete Frenet formula*

$$\Phi_{n+1}^m = \Phi_n^m L_n^m, \quad L_n^m = R(\kappa_{n+1}^m), \tag{8}$$

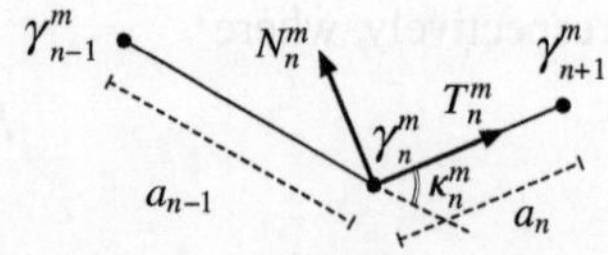

Fig. 1 Discrete plane curve

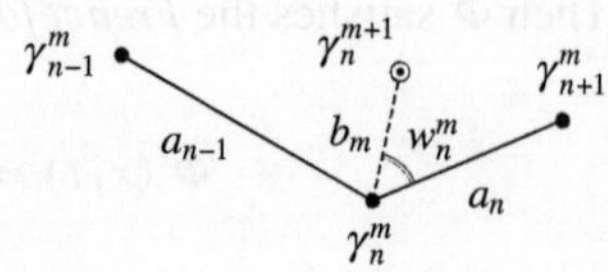

Fig. 2 Deformation of discrete plane curve

where κ_n^m is the angle between T_n^m and T_{n-1}^m (see Fig. 1). Now we assume the *isoperimetric condition*, namely, that $|\gamma_{n+1}^m - \gamma_n^m| = a_n$ is constant with respect to m, or

$$\left| \frac{\gamma_{n+1}^m - \gamma_n^m}{a_n} \right| = 1, \tag{9}$$

for all m. We also require the *equidistant condition*

$$\left| \frac{\gamma_n^{m+1} - \gamma_n^m}{b_m} \right| = 1, \tag{10}$$

for all n, where b_m is an arbitrary function in m. Then, putting the angle between the vectors $\gamma_{n+1}^m - \gamma_n$ and $\gamma_n^{m+1} - \gamma_n^m$ as w_n^m (see Fig. 2), the deformation of the discrete curve is expressed as

$$\frac{\gamma_n^{m+1} - \gamma_n^m}{b_m} = \cos w_n^m \, T_n^m + \sin w_n^m \, N_n^m. \tag{11}$$

The isoperimetric condition (9) and (11) implies that w_n^m is determined by the following equation

$$w_{n+1}^m = -\kappa_{n+1}^m + 2 \arctan \frac{b_m + a_n}{b_m - a_n} \tan \frac{w_n^m}{2}. \tag{12}$$

In terms of the discrete Frenet frame, the deformation is expressed as

$$\Phi_n^{m+1} = \Phi_n^m M_n^m, \quad M_n = R(\kappa_{n+1}^m + w_{n+1}^m + w_n^m). \tag{13}$$

The compatibility condition of the linear system (8) and (13) $L_n^m M_{n+1}^m = M_n^m L_n^{m+1}$ yields $w_{n+1}^m - w_{n-1}^m = \kappa_n^{m+1} - \kappa_{n+1}^m$ and the *discrete mKdV equation*

$$\frac{w_{n+1}^{m+1}}{2}-\frac{w_n^m}{2}=\arctan\left(\frac{b_{m+1}+a_n}{b_{m+1}-a_n}\tan\frac{w_n^{m+1}}{2}\right)-\arctan\left(\frac{b_m+a_{n+1}}{b_m-a_{n+1}}\tan\frac{w_{n+1}^m}{2}\right), \quad (14)$$

which is known as an integrable discretization of the mKdV equation [9, 20].

Remark 1 1. The continuous limit of the discrete mKdV equation (14) to the mKdV equation (1) is described as follows [16]. First put

$$a_n = a \text{ (const.)}, \quad b_m = b \text{ (const.)},$$
$$\delta = \frac{a+b}{2}, \quad \varepsilon = \frac{a-b}{2}, \quad \frac{s}{\delta} = n+m, \; l = n-m. \quad (15)$$

Taking the limit $\varepsilon \to 0$ yields the semi-discrete mKdV equation for $\kappa_l(s) = -w_l(s)$

$$\frac{d\kappa_l}{ds} = \frac{1}{\varepsilon}\left(\tan\frac{\kappa_{l+1}}{2} - \tan\frac{\kappa_{l-1}}{2}\right). \quad (16)$$

Next putting

$$x = \varepsilon l + s, \quad t = -\frac{\varepsilon^2}{6}s, \quad (17)$$

and taking the limit $\varepsilon \to 0$, we obtain the mKdV equation (1) for $\kappa(x, t)$. The semi-discrete mKdV equation (16) also describes certain isoperimetric flows on plane/space discrete curves [3, 16].

2. By using the standard technique of the theory of the integrable systems, it is possible to construct explicit formulas for γ in terms of the τ function [15, 16]. The τ functions corresponding to the soliton or the breather type solutions are expressed by determinants.

3 Space Curves

3.1 Isoperimetric Deformation Described by mKdV Equation

Let x be the arc-length, $\gamma(x, t) \in \mathbb{R}^3$ be an arc-length parameterized curve so that $|\gamma'| = 1$, and t be the deformation parameter. The Frenet frame is defined by

$$\Phi(x, t) = [T(x, t), N(x, t), B(x, t)] \in \mathrm{SO}(3),$$
$$T = \gamma', \; N = \frac{\gamma''}{|\gamma''|}, \quad B = T \times N, \quad (18)$$

where T, N, B are the *tangent vector*, the *normal vector* and the *binormal vector*, respectively. Then we have the *Frenet-Serret formula*

$$\Phi' = \Phi L, \quad L = \begin{bmatrix} 0 & -\kappa & 0 \\ \kappa & 0 & -\lambda \\ 0 & \lambda & 0 \end{bmatrix}, \ \kappa = |\gamma''|, \ \lambda = -\langle B', N\rangle, \tag{19}$$

where κ and λ are *curvature* and *torsion*, respectively, and $\langle \cdot, \cdot \rangle$ is the standard inner product. We assume that the torsion λ is a constant with respect to x, and consider the deformation of curves given by

$$\dot{\gamma} = \left(\frac{\kappa^2}{2} - 3\lambda^2\right) T + \kappa' N - 2\lambda\kappa B, \tag{20}$$

which can be expressed in terms of the Frenet frame as

$$\dot{\Phi} = \Phi M, \quad M = \begin{bmatrix} 0 & -\frac{\kappa^3}{2} + \lambda^2\kappa - \kappa'' & \lambda\kappa \\ \frac{\kappa^3}{2} - \lambda^2\kappa + \kappa'' & 0 & -\frac{\lambda}{2}\kappa^2 + \lambda^3 \\ -\lambda\kappa & \frac{\lambda}{2}\kappa^2 - \lambda^3 & 0 \end{bmatrix}. \tag{21}$$

The compatibility condition of the linear system (19) and (21) $\dot{L} - M' = [L, M]$ yields the mKdV equation (1). We remark that one can verify that this deformation preserves the arc-length and the torsion. Therefore, (20) defines an isoperimetric and torsion-preserving deformation of the smooth space curves with constant torsion described by the mKdV equation (1).

3.2 Isoperimetric Deformation of Discrete Space Curves

Now we introduce the discrete space curve, and formulate the isoperimetric, torsion-preserving and equidistant deformation of the space discrete curves with constant torsion. We refer to [17] for the details including the proof of the statements.

For $\gamma_n \in \mathbb{R}^3$, if any three consecutive points $\gamma_{n-1}, \gamma_n, \gamma_{n+1}$ are not colinear, we call γ_n a *discrete space curve*. We introduce the *discrete Frenet frame* by

$$\Phi_n = [T_n, N_n, B_n] \in \mathrm{SO}(3), \tag{22}$$

where the *tangent vector* T_n, the *normal vector* N_n and the *binormal vector* B_n are defined by

$$T_n = \frac{\gamma_{n+1} - \gamma_n}{\varepsilon_n}, \quad B_n = \frac{T_{n-1} \times T_n}{|T_{n-1} \times T_n|}, \quad N_n = B_n \times T_n, \tag{23}$$

respectively, where

$$\varepsilon_n = |\gamma_{n+1} - \gamma_n|. \tag{24}$$

Fig. 3 Discrete Frenet frame

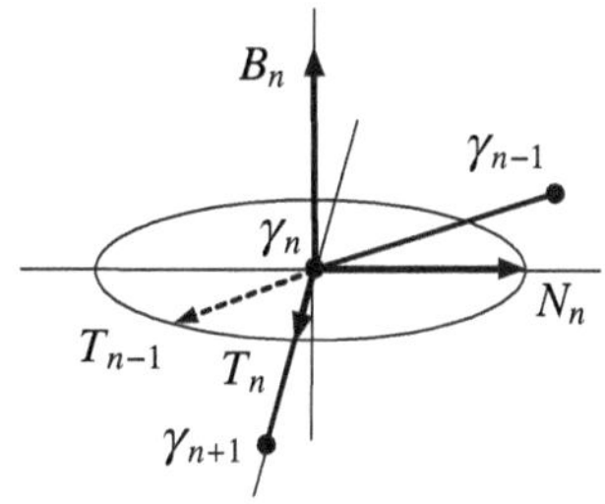

Fig. 4 κ_n and ν_n

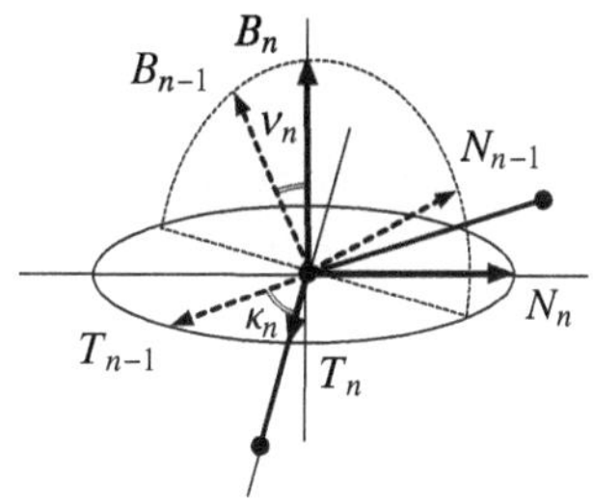

Note that the normal vector is chosen as $N_n \in \text{span}\{T_{n-1}, T_n\}$ (see Fig. 3). We sometimes call $\text{span}\{T_{n-1}, T_n\}$ the *osculating plane*.

We define $\kappa_n \in (0, \pi)$ and $\nu_n \in [-\pi, \pi)$ by (see Fig. 4)

$$\langle T_n, T_{n-1} \rangle = \cos \kappa_n, \quad \langle B_n, B_{n-1} \rangle = \cos \nu_n, \quad \langle B_n, N_{n-1} \rangle = \sin \nu_n. \tag{25}$$

Then we see that the Frenet frame satisfies the *discrete Frenet-Serret formula* [5]:

$$\Phi_{n+1} = \Phi_n L_n, \quad L_n = R_1(-\nu_{n+1}) R_3(\kappa_{n+1}), \tag{26}$$

where $R_1(\theta)$ and $R_3(\theta)$ are the rotation matrices given by

$$R_1(\theta) = \begin{bmatrix} 1 & 0 & 0 \\ 0 & \cos\theta & -\sin\theta \\ 0 & \sin\theta & \cos\theta \end{bmatrix}, \quad R_3(\theta) = \begin{bmatrix} \cos\theta & -\sin\theta & 0 \\ \sin\theta & \cos\theta & 0 \\ 0 & 0 & 1 \end{bmatrix}, \tag{27}$$

respectively. We define the function λ_n by

$$\lambda_n = \frac{\sin \nu_{n+1}}{\varepsilon_n}, \tag{28}$$

which we refer to as the *torsion* of γ_n. In the following, we assume that $\lambda_n = \lambda$ (const.). For later convenience we introduce a_n by

$$a_n = \left(1 + \tan^2 \frac{\nu_{n+1}}{2}\right) \varepsilon_n, \tag{29}$$

so that

$$\varepsilon_n = \frac{a_n}{1+\frac{a_n^2\lambda^2}{4}}, \quad \nu_{n+1} = 2\arctan\frac{a_n\lambda}{2}. \tag{30}$$

Now we formulate the discrete isoperimetric, torsion-preserving and equidistant deformation of the discrete space curve γ_n.

Theorem 1 *Let γ_n be a space discrete curve with constant torsion λ. We define the new discrete space curve $\overline{\gamma}_n$ by*

$$\overline{\gamma}_n = \gamma_n + \delta(\cos w_n T_n + \sin w_n N_n), \tag{31}$$

$$\delta = \frac{b}{1+\frac{b^2\lambda^2}{4}} > 0, \tag{32}$$

$$w_{n+1} = -\kappa_{n+1} + 2\arctan\left(\frac{b+a_n}{b-a_n}\tan\frac{w_n}{2}\right). \tag{33}$$

Here we choose the constant $b > 0$ and w_0 such that the sign of $\sigma_n = \sin(w_{n+1}+\kappa_{n+1}-w_{n-1})$ is constant for all n. Then we have the following:

1. *(Isoperimetricity) $\overline{\gamma}$ satisfies*

$$\overline{\varepsilon}_n = \left|\overline{\gamma}_{n+1} - \overline{\gamma}_n\right| = \left|\gamma_{n+1} - \gamma_n\right| = \varepsilon_n, \tag{34}$$

   and thus the deformation $\gamma_n \mapsto \overline{\gamma}_n$ defined by (31)–(33) *is an isoperimetric deformation.*
2. *(Preservation of the torsion) It follows that*

$$\overline{\nu}_n = \nu_n. \tag{35}$$

   Therefore, we have that $\overline{a}_n = a_n$, and that the torsion $\overline{\lambda}_n = \frac{\sin\overline{\nu}_{n+1}}{\overline{\varepsilon}_n}$ of $\overline{\gamma}_n$ is a constant with respect to n, whose value is equal to λ. Namely, the deformation $\gamma \mapsto \overline{\gamma}$ preserves the torsion.
3. *(Deformation of the Frenet frame) The Frenet frame $\overline{\Phi}_n = \left[\overline{T}_n, \overline{N}_n, \overline{B}_n\right]$ of $\overline{\gamma}_n$ satisfies either of the following: in case of $\sigma_n > 0$,*

$$\overline{\Phi}_n = \Phi_n R_3(w_n) R_1(\mu) R_3(w_{n+1}+\kappa_{n+1}), \quad \mu = -2\arctan\frac{b\lambda}{2}, \tag{36}$$

   or in case of $\sigma_n < 0$,

$$\overline{\Phi}_n = \Phi_n R_3(w_n) R_1(\mu) R_3(-w_{n+1}-\kappa_{n+1}), \quad \mu = 2\arctan\frac{2}{b\lambda}. \tag{37}$$

Repeating the construction in Theorem 1 yields the sequence of the discrete space curves with constant torsion $\gamma_n^0 = \gamma_n, \gamma_n^1 = \overline{\gamma}_n, \gamma_n^2 = \overline{\gamma_n^1}, \ldots, \gamma_n^m = \overline{\gamma_n^{m-1}}, \ldots$. Correspondingly, we write data of the discrete curves as κ_n^m, w_n^m and so forth. Noticing that one may change the constant b at each step, the deformation of the curve is

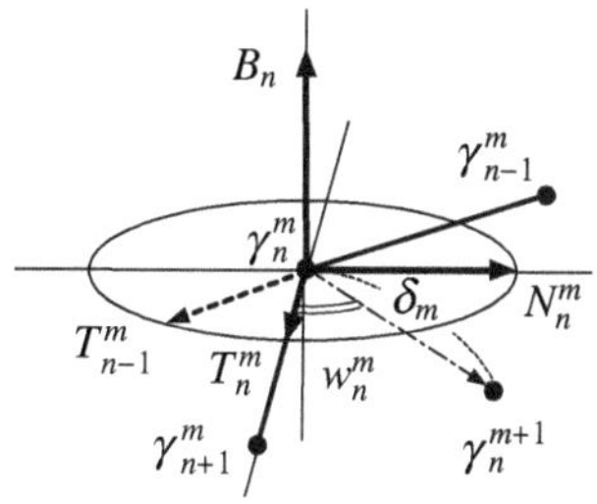

Fig. 5 Deformation of curve

rewritten as (see Fig. 5)

$$\gamma_n^{m+1} = \gamma_n^m + \delta_m \left(\cos w_n^m T_n^m + \sin w_n^m N_n^m\right), \quad \delta_m = \frac{b_m}{1 + \frac{b_m^2 \lambda^2}{4}}, \tag{38}$$

$$w_{n+1}^m = -\kappa_{n+1}^m + 2 \arctan \frac{b_m + a_n}{b_m - a_n} \tan \frac{w_n^m}{2}, \tag{39}$$

where b_m is an arbitrary function in m and the sign of $\sigma_n^m = \sin(w_{n+1}^m + \kappa_{n+1}^m - w_{n-1}^m)$ should be constant with respect to n. (38) and (39) define an isoperimetric, torsion-preserving and equidistant deformation of the discrete space curves. The Frenet frame Φ_n^m satisfies

$$\Phi_{n+1}^m = \Phi_n^m L_n^m, \quad \Phi_n^{m+1} = \Phi_n^m M_n^m. \tag{40}$$

Here,

$$L_n^m = R_1\left(-\nu_{n+1}\right) R_3\left(\kappa_{n+1}^m\right), \quad \nu_{n+1} = 2 \arctan \frac{a_n \lambda}{2}, \tag{41}$$

and M_n^m is given by either of the following for each m: in case of $\sigma_n^m > 0$,

$$M_n^m = R_3\left(w_n^m\right) R_1\left(\mu_m\right) R_3\left(w_{n+1}^m + \kappa_{n+1}^m\right), \quad \mu_m = -2 \arctan \frac{b_m \lambda}{2}, \tag{42}$$

or in case of $\sigma_n^m < 0$,

$$M_n^m = R_3\left(w_n^m\right) R_1\left(\mu_m\right) R_3\left(-w_{n+1}^m - \kappa_{n+1}^m\right), \quad \mu_m = 2 \arctan \frac{2}{b_m \lambda}. \tag{43}$$

Choosing the matrix M as (42), the compatibility condition $L_n^m M_{n+1}^m = M_n^m L_n^{m+1}$ for (40) gives

$$w_{n+1}^m - w_{n-1}^m = \kappa_n^{m+1} - \kappa_{n+1}^m. \tag{44}$$

Then (39) and (44) yields the discrete mKdV equation

$$\frac{w_{n+1}^{m+1}}{2} - \frac{w_n^m}{2} = \arctan\left(\frac{b_{m+1} + a_n}{b_{m+1} - a_n} \tan \frac{w_n^{m+1}}{2}\right) - \arctan\left(\frac{b_m + a_{n+1}}{b_m - a_{n+1}} \tan \frac{w_{n+1}^m}{2}\right). \tag{45}$$

Choosing the matrix M as (43), the compatibility condition of (40) gives

$$w_{n+1}^{m} - w_{n-1}^{m} = -\kappa_{n}^{m+1} - \kappa_{n+1}^{m}, \tag{46}$$

and then (39) and (46) yields the *discrete sine-Gordon equation*

$$\frac{w_{n+1}^{m+1}}{2} + \frac{w_n^m}{2} = \arctan\left(\frac{b_{m+1}+a_n}{b_{m+1}-a_n}\tan\frac{w_n^{m+1}}{2}\right) + \arctan\left(\frac{b_m + a_{n+1}}{b_m - a_{n+1}}\tan\frac{w_{n+1}^m}{2}\right). \tag{47}$$

Remark 2 1. One can see from (30) and (38) that the lattice parameters ε, δ for the space discrete curve and a, b for the discrete mKdV or the discrete sine-Gordon equations are different. This is an essential feature of the discrete deformation of space discrete curves. They coincide by setting $\lambda = 0$ (case of plane discrete curves).
2. The discrete sine-Gordon equation (47) can be rewritten as the well-known form [8]

$$\sin\frac{\theta_{n+1}^{m+1} - \theta_n^{m+1} - \theta_{n+1}^m + \theta_n^m}{4} = \frac{a_n}{b_m}\sin\frac{\theta_{n+1}^{m+1} + \theta_n^{m+1} + \theta_{n+1}^m + \theta_n^m}{4}, \tag{48}$$

in terms of the potential function defined by $w_n^m = -\frac{\theta_n^{m+1} + \theta_{n+1}^m}{2}$.

It is the necessary and sufficient condition for the deformation of the discrete curve $\gamma_n^m \mapsto \gamma_n^{m+1}$ being torsion-preserving that the sign of σ_n^m is constant as a function of n at each m. Moreover, the deformation of the Frenet frame is different according to the sign of σ_n^m. It is possible to control the sign of σ_n^m and thus the deformation of the Frenet frame by tuning the deformation parameter b_m according to the data of γ_n^m as follows:

Proposition 1 1. *If we choose $b_m > 0$ to satisfy either of the following conditions at each m, then the deformation defined by* (38) *and* (39) *is torsion-preserving.*

$$\text{(i)}\quad b_m > \max\left\{\frac{a_{\max}}{\tan\frac{\kappa_{\min}^m}{4}}, \frac{\Delta a}{2\cos\frac{\kappa_{\max}^m}{2}}\left(1 + \sqrt{1 + \frac{4a_{\min}a_{\max}}{(\Delta a)^2}\cos^2\frac{\kappa_{\max}^m}{2}}\right)\right\},$$

or (49)

$$\text{(ii)}\quad b_m < \min\left\{a_{\min}\tan\frac{\kappa_{\min}^m}{4}, \frac{\Delta a}{2\cos\frac{\kappa_{\max}^m}{2}}\left(-1 + \sqrt{1 + \frac{4a_{\min}a_{\max}}{(\Delta a)^2}\cos^2\frac{\kappa_{\max}^m}{2}}\right)\right\},$$

where

$$\begin{gathered}\kappa_{\min}^m = \min_n \kappa_n^m, \quad \kappa_{\max}^m = \max_n \kappa_n^m, \\ a_{\max} = \max_n a_n, \quad a_{\min} = \min_n a_n, \quad \Delta a = a_{\max} - a_{\min}.\end{gathered} \tag{50}$$

2. *In the case of* (i), *it holds that $\sigma_n^m < 0$ and thus the Frenet frame Φ is deformed according to* (40), (41), (43). *In the case of* (ii), *it holds that $\sigma_n^m > 0$ and thus Φ is deformed according to* (40), (41), (42). *Namely,* (i) *and* (ii) *correspond to the deformation by the discrete sine-Gordon equation* (47) *and discrete mKdV equation* (45), *respectively.*

Remark 3 1. For given constant λ and functions $a_n, b_m, w_n^m, \kappa_n^m$, it is possible to reconstruct the discrete curves γ_n^m as follows: let Φ_n^m be a solution to the linear system (40), (41), (42) (or (43)). Transforming Φ_n^m to SU(2)-valued function ϕ_n^m by the SU(2)-SO(3) correspondence, the *Sym-Tafel formula* recovers γ_n^m:

$$\gamma_n^m = -\left(\frac{\partial}{\partial\lambda}\phi_n^m\right)(\phi_n^m)^{-1} \in \mathfrak{su}(2) \simeq \mathbb{R}^3. \tag{51}$$

2. The sequence of isoperimetric deformations of space discrete curves γ_n^m form discrete surfaces with the constant negative Gaussian curvature (discrete K-surfaces) [1]. Conversely, any discrete K-surface can be constructed by γ_n^m.

References

1. Bobenko A, Pinkall U (1996) Discrete surfaces with constant negative Gaussian curvature and the Hirota equation. J Diff Geom 43:527–611
2. Doliwa A, Santini PM (1994) An elementary geometric characterization of the integrable motions of a curve. Phys Lett A 185:373–384
3. Doliwa A, Santini PM (1995) Integrable dynamics of a discrete curve and the Ablowitz-Ladik hierarchy. J Math Phys 36:1259–1273
4. Doliwa A, Santini PM (1996) The integrable dynamics of a discrete curve. In: Levi D, Vinet L, Winternitz P (eds) Symmetries and integrability of difference equations. CRM proceedings and lecture notes, vol 9. American Mathematical Society, Providence, RI, pp 91–102
5. Eyring H (1932) The resultant electric moment of complex molecules. Phys Rev 39:746–748
6. Goldstein RE, Petrich DM (1991) The Korteweg-de Vries hierarchy as dynamics of closed curves in the plane. Phys Rev Lett 67:3203–3206
7. Hasimoto H (1972) A soliton on a vortex filament. J Fluid Mech 51:477–485
8. Hirota R (1977) Nonlinear partial difference equations III; discrete sine-Gordon equation. J Phys Soc Jpn 43:2079–2086
9. Hirota R (1998) Discretization of the potential modified KdV equation. J Phys Soc Jpn 67:2234–2236
10. Hisakado M, Nakayama K, Wadati M (1995) Motion of discrete curves in the plane. J Phys Soc Jpn 64:2390–2393
11. Hisakado M, Wadati M (1996) Moving discrete curve and geometric phase. Phys Lett A 214:252–258
12. Hoffmann T (2008) Discrete Hashimoto surfaces and a doubly discrete smoke-ring flow. In: Bobenko AI, Schröder P, Sullivan JM, Ziegler GM (eds) Discrete differential geometry. Oberwolfach seminars, vol 38. Birkhäuser, Basel, pp 95–115
13. Hoffmann T (2009) Discrete differential geometry of curves and surfaces. COE lecture notes, vol 18. Kyushu University, Fukuoka
14. Hoffmann T, Kutz N (2004) Discrete curves in $\mathbb{C}P^1$ and the Toda lattice. Stud Appl Math 113:31–55
15. Inoguchi J, Kajiwara K, Matsuura N, Ohta Y (2012) Motion and Bäcklund transformations of discrete plane curves. Kyushu J Math 66:303–324
16. Inoguchi J, Kajiwara K, Matsuura N, Ohta Y (2012) Explicit solutions to the semi-discrete modified KdV equation and motion of discrete plane curves. J Phys A: Math Theor 45:045206
17. Inoguchi J, Kajiwara K, Matsuura N, Ohta Y (preprint) Discrete mKdV and discrete sine-Gordon flows on discrete space curves. arXiv:1311.4245
18. Lamb G Jr (1976) Solitons and the motion of helical curves. Phys Rev Lett 37:235–237

19. Langer J, Perline R (1998) Curve motion inducing modified Korteweg-de Vries systems. Phys Lett A 239:36–40
20. Matsuura N (2012) Discrete KdV and discrete modified KdV equations arising from motions of planar discrete curves. Int Math Res Not 2012:1681–1698
21. Nakayama K (2007) Elementary vortex filament model of the discrete nonlinear Schrödinger equation. J Phys Soc Jpn 76:074003
22. Nakayama K, Segur H, Wadati M (1992) Integrability and the motions of curves. Phys Rev Lett 69:2603–2606
23. Nishinari K (1999) A discrete model of an extensible string in three-dimensional space. J Appl Mech 66:695–701
24. Pinkall U, Springborn B, Weißmann S (2007) A new doubly discrete analogue of smoke ring flow and the real time simulation of fluid flow. J Phys A: Math Theor 40:12563–12576

Mathematical Formulation of Motion and Deformation and Its Applications

Hiroyuki Ochiai and Ken Anjyo

Abstract This chapter is intended to give a summary of Lie groups and Lie algebras for computer graphics, including an example from interpolations and blending of motions and deformation. In animation and filmmaking procedure, we want to get a smooth transition of the drawing of given starting and ending point (the drawings at these points are called *keyframe*). This problem can be understood as an *interpolation*, so that various approach have been proposed and used in Computer Graphics. After the seminal work so-called ARAP (as-rigid-as-possible), serious properties of matrix groups have been taken into account both in theoretical and in computational point of view. To understand and develop these properties, we here employ a Lie theoretic approach.

Keywords Blend shape · Lie group · Lie algebra · Polar decomposition · As-rigid-as-possible · Deformation · Exponential map · Motion group

1 Lie Groups and Lie Algebras

Lie groups are used for describing the transformations, and Lie algebra is its *linear* approximation which is a main feature of Lie groups.

A *Lie group* is a manifold with a group structure. A typical example of a Lie group is a subset of a matrix space $M(N) = M(N, \mathbb{R})$ for some N closed under the product

H. Ochiai (✉)
Institute of Mathematics for Industry, Kyushu University/JST CREST, 744, Motooka, Nishi-ku, Fukuoka 819-0395, Japan
e-mail: ochiai@imi.kyushu-u.ac.jp

K. Anjyo
OLM Digital, Inc./JST CREST, Dens Kono Bldg. 302, 1-8-8, Wakabayashi, Setagaya-ku, Tokyo 154-0023, Japan
e-mail: anjyo@olm.co.jp

K. Anjyo (ed.), *Mathematical Progress in Expressive Image Synthesis I*, Mathematics for Industry 4, DOI: 10.1007/978-4-431-55007-5_16,

and the inverse. To focus on applications to computer graphics, we may restrict Lie groups to be of this sub-class without loss of generality.

An affine transformation of $\mathbb{R}^n$ is a map from $\mathbb{R}^n$ to $\mathbb{R}^n$ which maps every line to a line. We denote by $\mathrm{Aff}(n)$ the set of affine transformations, and denote by $\mathrm{Aff}^+(n)$ the set of positive (i.e., reflection-free) affine transformations. In computer graphics, we mainly treat $n = 2$ and $n = 3$, while a part of the theory holds for a general n in a parallel manner. We see that $\mathrm{Aff}(n)$ is a group and $\mathrm{Aff}^+(n)$ is a subgroup. A well-known homogeneous realization of $\mathrm{Aff}(n)$ is given as a set of block upper-triangular matrices:

$$\mathrm{Aff}(n) = \left\{ A = \begin{pmatrix} \hat{A} & d_A \\ 0 & 1 \end{pmatrix} \middle| \, \hat{A} \in GL(n), d_A \in \mathbb{R}^n \right\}, \tag{1}$$

where $GL(n) = \{A \in M(n, \mathbb{R}) \mid \det(A) \neq 0\}$. This realization shows that $\mathrm{Aff}(n)$ is a Lie group.

The exponential of a square matrix $X \in M(N)$ is defined to be

$$\exp(X) = \sum_{k=0}^{\infty} \frac{1}{k!} X^k, \tag{2}$$

motivated by the exactly the same Taylor expansion formula for the scalar exponential function. This infinite series always converges; however, the computation is reduced to the exponential of diagonal matrices by using the formula $\exp(PXP^{-1}) = P \exp(X) P^{-1}$.

Let G be a Lie group realized in $M(N, \mathbb{R})$. The tangent space of G at the origin is denoted by $\mathfrak{g}$, the corresponding German letter, and is called the *Lie algebra* of G. Rather surprisingly, the Lie algebra approximates the Lie group, without losing the (local) informations. The exponential map gives a map from $\mathfrak{g}$ to G. If G is abelian (i.e., commutative) and connected, then the exponential map is surjective. If G is connected, simply connected, and abelian, then the exponential map is bijective. If G is compact and connected, then the exponential map is surjective. The exponential map is a local isomorphism at the origin, but it is not necessarily injective or surjective, e.g. $G = SL(2, \mathbb{R})$. To mitigate this difficulty on the exponential functions for general Lie groups, we will introduce the decomposition/factorization of matrices.

2 Matrix Factorization

For further understanding and analysis of complicated groups, several types decompositions of Lie groups and their generalization are known and used. We here review some of them (see also [5]).

2.1 Semi-direct Product Group

The affine transformation group is an example of a semi-direct product group, so $\mathrm{Aff}(n) = GL(n) \ltimes \mathbb{R}^n$. In general, let G be a group and H_1, H_2 be subgroups of G. If the multiplication map

$$H_1 \times H_2 \ni (h_1, h_2) \mapsto h_1 h_2 \in G \tag{3}$$

is bijective and $h_1 h_2 h_1^{-1} \in H_2$ for all $h_1 \in H_1$ and $h_2 \in H_2$, then G is isomorphic to the semi-direct product group $H_1 \ltimes H_2$. Another example of the semi-direct product group is a motion group $SE^+(n) = SO(n) \ltimes \mathbb{R}^n$ or a congruence group $SE(n) = O(n) \ltimes \mathbb{R}^n$. If $h_1 h_2 h_1^{-1} = h_2$ in the above, then G is isomorphic to the direct product group $H_1 \times H_2$.

For $\mathbb{K} = \mathbb{R}, \mathbb{C}$, or $\mathbb{H}$, the group of invertible dual numbers (see [4] Definition 1 for the notation) is

$$\hat{\mathbb{K}}^\times = (\mathbb{K} + \mathbb{K}\varepsilon)^\times = \mathbb{K}^\times + \mathbb{K}\varepsilon = \mathbb{K}^\times \ltimes \mathbb{K}\varepsilon, \tag{4}$$

which also gives an example of semi-direct product groups. The set $\hat{\mathbb{K}}_1$ of unit dual numbers is a subgroup of $\hat{\mathbb{K}}^\times$, which is also a semi-direct product group $\mathbb{K}_1 \ltimes \mathbb{K}\varepsilon$.

2.2 Diagonalization of Symmetric Matrix

Every (real) symmetric matrix is diagonalized by the conjugate of an orthogonal matrix. This fact is rephrased as the subjectivity of the multiplication map

$$O(n) \times \mathrm{Diag}(n) \ni (R, D) \mapsto RDR^{-1} \in \mathrm{Sym}(n). \tag{5}$$

Note that the diagonal entries of diagonal matrix D are the set of eigenvalues (with multiplicities) of a given symmetric matrix $X = RDR^{-1}$, so it is determined by X up to the ordering of eigenvalues. The column vectors of R are the corresponding eigenvectors, which form orthonormal frame of $\mathbb{R}^n$. These facts give an algorithm to compute (R, D) from X. As a special case of this decomposition (5), every positive-definite symmetric matrix is diagonalized into diagonal matrices with positive diagonal entries:

$$O(n) \times \mathrm{Diag}^+(n) \ni (R, D) \mapsto RDR^{-1} \in \mathrm{Sym}^+(n). \tag{6}$$

Note that $\mathrm{Sym}^+(n)$ is not a group since the product of two symmetric matrices is not necessarily symmetric. But the exponential map still gives a bijective from $\mathrm{Sym}(n)$ to $\mathrm{Sym}^+(n)$. This enables us to consider the logarithm, which is defined to be the inverse of the exponential, and the fractional power S^t for $t \in \mathbb{R}$ defined to be

$\exp(t \log(S))$ for $S \in \mathrm{Sym}^+(n)$. (An example is the square root $S^{1/2}$.) This is a key for interpolations.

2.3 Polar Decomposition

Every invertible matrix is the product of an orthogonal matrix and a positive-definite symmetric matrix. In other words, the multiplication map

$$O(n) \times \mathrm{Sym}^+(n) \ni (R, S) \mapsto RS \in GL(n) \tag{7}$$

is bijective.

2.4 Singular Value Decomposition (SVD)

The following multiplication map is surjective:

$$SO(n) \times \mathrm{Diag}^+(n) \times SO(n) \ni (R, D, R') \mapsto RDR' \in GL^+(n). \tag{8}$$

2.5 Triangular Decomposition

In some settings, almost all matrices can be decomposed. Here is an example from Gaussian elimination of systems of linear equations. Let $N^{\pm}$ be the set of upper/lower triangular matrices whose diagonal entries are 1. Then the multiplication map

$$N^- \times \mathrm{Diag}(n) \times N^+ \ni (U', D, U) \mapsto U'DU \in GL(n) \tag{9}$$

is injective and its image is an open dense subset of $GL(n)$.

2.6 Iwasawa Decomposition

Some decomposition can treat shears. The multiplication map

$$SO(n) \times \mathrm{Diag}^+(n) \times N^+ \ni (R, D, U) \mapsto RDU \in GL^+(n) \tag{10}$$

gives a bijection.

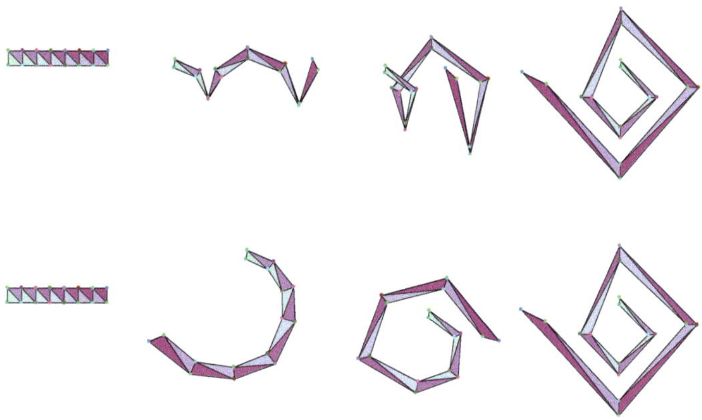

Fig. 1 2D deformation, comparison of [1] and [3]

3 Interpolation and Blend

Our goal is to give an interpolation on $GL^+(2)$, but this space is not a convex subset of $M(2)$. By combining the matrix factorization and the exponential map, we get a straightforward idea for interpolation. Graphically,

$$\begin{array}{c} GL^+(2) \overset{\sim}{\to} SO(2) \times \mathrm{Sym}^+(2) \overset{\log}{\to} \mathfrak{so}(2) \times \mathfrak{sym}(2) \\ \downarrow \text{linear interpolation} \\ GL^+(2) \overset{\sim}{\leftarrow} SO(2) \times \mathrm{Sym}^+(2) \overset{\exp}{\leftarrow} \mathfrak{so}(2) \times \mathfrak{sym}(2). \end{array} \tag{11}$$

By decomposition, we switch to the spaces whose exponential maps behave well, and then by the exponential map, we move to a linear space where we have a reasonable interpolation. This is a modification of the idea in [1], used in [3] (See Fig. 1).

Another application of this idea is

$$\begin{array}{c} \mathrm{Aff}^+(3) \overset{\sim}{\to} SE(3) \times \mathrm{Sym}^+(3) \overset{\log}{\to} \mathfrak{se}(3) \times \mathfrak{sym}(3) \\ \downarrow \text{linear interpolation} \\ \mathrm{Aff}^+(3) \overset{\sim}{\leftarrow} SE(3) \times \mathrm{Sym}^+(3) \overset{\exp}{\leftarrow} \mathfrak{se}(3) \times \mathfrak{sym}(3), \end{array} \tag{12}$$

where $SE(3)$ is a three-dimensional Euclidian motion group with its Lie algebra $\mathfrak{se}(3)$ (See Fig. 2).

The exponential $\exp \begin{pmatrix} X & d \\ 0 & 0 \end{pmatrix} = \begin{pmatrix} R & l \\ 0 & 1 \end{pmatrix}$ of an element $\mathfrak{se}(3)$ has a Rodrigues' type formula, with $2\cos\theta = \mathrm{Tr}(R)$, or $-2\theta^2 = \mathrm{Tr}(X)$,

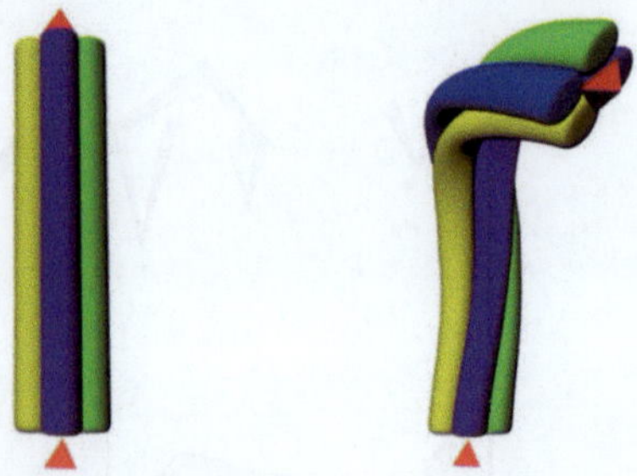

Fig. 2 Probe-based deformation

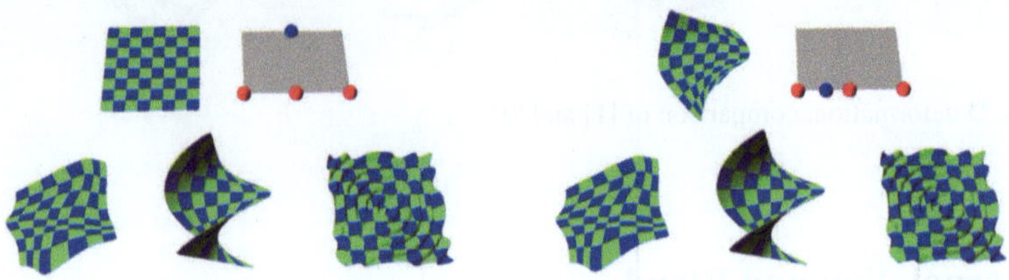

Fig. 3 Blend with arbitrary weights

Fig. 4 Deformation by using anti-commutative DCN

$$X = \frac{\theta}{2\sin\theta}(R - {}^tR), \text{ and } l = \left(I - \frac{1}{2}X + \frac{2\sin\theta - (1+\cos\theta)\theta}{2\theta^2\sin\theta}X^2\right)d. \quad (13)$$

Note that the right-most column of (11) and (12) is a vector space, so we can blend two or more objects just as a linear combination $\sum_{i=1}^{m} w_i S_i$ with weights w_i. Further examples are given in [2] (See Fig. 3).

Figure 4 shows a deformation by using 'anti-commutative dual complex numbers' [4].

Acknowledgments This work was supported by Core Research for Evolutional Science and Technology (CREST) Program "Mathematics for Computer Graphics" of Japan Science and Technology Agency (JST). The authors are grateful to the collaborators including Shizuo Kaji at Yamaguchi University, Yoshihiro Mizoguchi, Shun'ichi Yokoyama, Hiroyasu Hamada, and Kohei Matsushita, Genki Matsuda at Kyushu University and Ayumi Kimura, Sampei Hirose, and Gengdai Liu at OLM Digital for their valuable discussions.

References

1. Alexa M, Cohen-Or D, Levin D (2000) As-rigid-as-possible shape interpolation. In: Proceedings of the 27th annual conference on computer graphics and interactive techniques. SIGGRAPH 2000, pp. 157–164
2. Kaji S, Hirose S, Ochiai H, Anjyo K (2013) A Lie theoretic parameterization of affine transformation. In: Mathematical progress in expressive image synthesis, MI Lecture Note, vol 50. Kyushu University, pp 134–140
3. Kaji S, Hirose S, Sakata S, Mizoguchi Y, Anjyo K (2012) Mathematical analysis on affine maps for 2D shape interpolation. In: Proceedings of SCA2012, pp 71–76
4. Matsuda G, Kaji S, Ochiai H (2013) Anti-commutative dual complex numbers and 2D rigid transformation. In: Mathematical progress in expressive image synthesis, MI Lecture Note, vol 50. Kyushu University, pp 128–133
5. Ochiai H, Anjyo K (2013) Mathematical description of motion and deformation—from basics to graphics applications. SIGGRAPH Asia 2013 course. http://portal.acm.org. (Revised course notes are also available at http://mcg.imi.kyushu-u.ac.jp/english/index.php)

Anti-commutative Dual Complex Numbers and 2D Rigid Transformation

Genki Matsuda, Shizuo Kaji and Hiroyuki Ochiai

Abstract We introduce a new presentation of the two dimensional rigid transformation which is more concise and efficient than the standard matrix presentation. By modifying the ordinary dual number construction for the complex numbers, we define the ring of *anti-commutative dual complex numbers*, which parametrizes two dimensional rotation and translation all together. With this presentation, one can easily interpolate or blend two or more rigid transformations at a low computational cost. We developed a library for C++ with the MIT-licensed source code [13].

Keywords 2D deformation · 2D animation · Interpolation · Skinning · Rigid transformation · Euclidian motion · Dual number

1 Rigid Transformation

The n-dimensional *rigid transformation* (*or Euclidean*) group $\mathrm{E}(n)$ consists of transformations of $\mathbb{R}^n$ which preserves the standard metric. This group serves as an essential mathematical backend for many applications (see [9]). It is well-known (see [5], for example) that any element of $\mathrm{E}(n)$ can be written as a composition of

G. Matsuda
Graduate School of Mathematics, Kyushu University,
744, Motooka, Nishi-ku, Fukuoka 819-0395, Japan
e-mail: ma212041@math.kyushu-u.ac.jp

S. Kaji (✉)
Yamaguchi University/JST CREST, 1677-1, Yoshida, Yamaguchi 753-8512, Japan
e-mail: skaji@yamaguchi-u.ac.jp

H. Ochiai
Institute of Mathematics for Industry, Kyushu University/JST CREST, 744,
Motooka, Nishi-ku, Fukuoka 819-0395, Japan
e-mail: ochiai@imi.kyushu-u.ac.jp

K. Anjyo (ed.), *Mathematical Progress in Expressive Image Synthesis I*,
Mathematics for Industry 4, DOI: 10.1007/978-4-431-55007-5_17,

a rotation, a reflection, and a translation, and hence, it is represented by $(n+1) \times (n+1)$-homogeneous matrix;

$$\mathrm{E}(n) = \left\{ A = \begin{pmatrix} \hat{A} & d_A \\ 0 & 1 \end{pmatrix} \mid \hat{A}^t \hat{A} = I_n, d_A \in \mathbb{R}^n \right\}.$$

Here, we adopt the convention that a matrix acts on a (column) vector by the multiplication from the left. $\mathrm{E}(n)$ has two connected components. The identity component $\mathrm{SE}(n)$ consists of those without reflection. More precisely,

$$\mathrm{SE}(n) = \left\{ A = \begin{pmatrix} \hat{A} & d_A \\ 0 & 1 \end{pmatrix} \mid \hat{A} \in \mathrm{SO}(n), d_A \in \mathbb{R}^n \right\},$$

where $\mathrm{SO}(n) = \{R \mid R^t R = I_n, \det(R) = 1\}$ is the *special orthogonal group* composed of n-dimensional rotations.

The group $\mathrm{SE}(n)$ is widely used in computer graphics such as for expressing motion and attitude, displacement [10], deformation [1, 4, 11], skinning [7], and camera control [2]. In some cases, the matrix representation of the group $\mathrm{SE}(n)$ is not convenient. In particular, a linear combination of two matrices in $\mathrm{SE}(n)$ does not necessarily belong to $\mathrm{SE}(n)$ and it causes the notorious *candy-wrapper* defect in skinning. When $n = 3$, another representation of SE(3) using the *dual quaternion numbers* (DQN, for short) is considered in [7] to solve the candy-wrapper defect. In this chapter, we consider the 2-dimensional case. Of course, 2D case can be handled by regarding the plane embedded in $\mathbb{R}^3$ and using DQN, but instead, we introduce the *anti-commutative dual complex numbers* (DCN, for short), which is specific to 2D with much more concise and faster implementation [13]. To summarise, our DCN has the following advantages:

- any number of rigid transformations can be blended/interpolated easily using its algebraic structure with no degeneration defects such as the candy-wrapper defect (see Sect. 5)
- it is efficient in terms of both memory and CPU usage (see Sect. 6).

We believe that our DCN offers a choice for representing the 2D rigid transformation in certain applications which requires the above properties.

2 Anti-commutative Dual Complex Numbers

Let $\mathbb{K}$ denote one of the fields $\mathbb{R}$, $\mathbb{C}$, or $\mathbb{H}$, where $\mathbb{H}$ is the quaternion numbers. First, we recall the standard construction of the *dual numbers* over $\mathbb{K}$.

Definition 1 The ring of dual numbers $\hat{\mathbb{K}}$ is the quotient ring defined by

$$\hat{\mathbb{K}} := \mathbb{K}[\varepsilon]/(\varepsilon^2) = \{p_0 + p_1\varepsilon \mid p_0, p_1 \in \mathbb{K}\}.$$

We often denote an element in $\hat{\mathbb{K}}$ by a symbol with hat such as $\hat{p}$.

The addition and the multiplication of two dual numbers are given as

$$(p_0 + p_1\varepsilon) + (q_0 + q_1\varepsilon) = (p_0 + q_0) + (p_1 + q_1)\varepsilon,$$
$$(p_0 + p_1\varepsilon)(q_0 + q_1\varepsilon) = (p_0q_0) + (p_1q_0 + p_0q_1)\varepsilon.$$

The following involution is considered to be the dual version of conjugation

$$\widetilde{p_0 + p_1\varepsilon} := p_0^* - p_1^*\varepsilon,$$

where p_i^* is the usual conjugation of p_i in $\mathbb{K}$. (Note that in some literatures $\tilde{\hat{p}}$ is denoted by $\overline{\hat{p}^*}$.)

The *unit dual numbers* are of special importance.

Definition 2 Let $|\hat{p}| = \sqrt{\hat{p}\tilde{\hat{p}}}$ for $\hat{p} \in \hat{\mathbb{K}}$. The unit dual numbers is defined as

$$\hat{\mathbb{K}}_1 := \{\hat{p} \in \hat{\mathbb{K}} \mid |\hat{p}| = 1\} \subset \hat{\mathbb{K}}.$$

$\hat{\mathbb{K}}_1$ acts on $\hat{\mathbb{K}}$ by conjugation action

$$\hat{p} \diamond \hat{q} := \hat{p}\hat{q}\tilde{\hat{p}}$$

where $\hat{p} \in \hat{\mathbb{K}}_1, \hat{q} \in \hat{\mathbb{K}}$.

The unit dual quaternion $\hat{\mathbb{H}}_1$ is successfully used for skinning in [7]; a vector $v = (x, y, z) \in \mathbb{R}^3$ is identified with $1 + (xi + yj + zk)\varepsilon \in \hat{\mathbb{H}}_1$ and the conjugation action of $\hat{\mathbb{H}}_1$ preserves the embedded $\mathbb{R}^3$ and its Euclidean metric. In fact, the conjugation action induces the double cover $\hat{\mathbb{H}}_1 \to \mathrm{SE}(3)$.

On the other hand, when $\mathbb{K} = \mathbb{R}$ or $\mathbb{C}$, the conjugation action is trivial since the corresponding dual numbers are commutative. Therefore, we define the *anti-commutative dual complex numbers* (DCN, for short) $\check{\mathbb{C}}$ by modifying the multiplication of $\hat{\mathbb{C}}$. That is, $\check{\mathbb{C}} = \hat{\mathbb{C}}$ as a set, and the algebraic operations are replaced by

$$(p_0 + p_1\varepsilon)(q_0 + q_1\varepsilon) = (p_0q_0) + (p_1\tilde{q}_0 + p_0q_1)\varepsilon,$$
$$\widetilde{p_0 + p_1\varepsilon} = \tilde{p}_0 + p_1\varepsilon,$$
$$|p_0 + p_1\varepsilon| = |p_0|.$$

The addition and the conjugation action are kept unchanged. Then,

Theorem 1 $\check{\mathbb{C}}$ *satisfies distributive and associative laws, and hence, has a (non-commutative) ring structure.*

Similarly to the unit dual quaternion numbers, the unit anti-commutative complex numbers are of particular importance:

$$\check{\mathbb{C}}_1 := \{\hat{p} \in \check{\mathbb{C}} \mid |\hat{p}| = 1\} = \{e^{i\theta} + p_1\varepsilon \in \check{\mathbb{C}} \mid \theta \in \mathbb{R}, p_1 \in \mathbb{C}\}.$$

It forms a group with inverse

$$(e^{i\theta} + p_1\varepsilon)^{-1} = e^{-i\theta} - p_1\varepsilon.$$

We define an action of $\check{\mathbb{C}}_1$ on $\check{\mathbb{C}}$ by the conjugation. Now, we regard $\mathbb{C} = \mathbb{R}^2$ as usual. Identifying $v \in \mathbb{C}$ with $1 + v\varepsilon \in \check{\mathbb{C}}$, we see that $\check{\mathbb{C}}_1$ acts on $\mathbb{C}$ as rigid transformation.

3 Relation to SE(2)

In the previous section, we constructed the unit anti-commutative dual complex numbers $\check{\mathbb{C}}_1$ and its action on $\mathbb{C} = \mathbb{R}^2$ as rigid transformation. Recall that $\hat{p} = p_0 + p_1\varepsilon \in \check{\mathbb{C}}_1$ acts on $v \in \mathbb{C}$ by

$$\hat{p} \diamond (1 + v\varepsilon) = (p_0 + p_1\varepsilon)(1 + v\varepsilon)(\tilde{p}_0 + p_1\varepsilon) = 1 + (p_0^2 v + 2p_0p_1)\varepsilon, \qquad (1)$$

that is, v maps to $p_0^2 v + 2p_0p_1$. For example, when $p_1 = 0$, $v \in \mathbb{C}$ is mapped to $p_0^2 v$, which is the rotation around the origin of degree $2\arg(p_0)$ since $|p_0| = 1$. On the other hand, when $p_0 = 1$, the action corresponds to the translation by $2p_1$. In general, we have

$$\varphi : \check{\mathbb{C}}_1 \to \mathrm{SE}(2)$$
$$p_0 + p_1\varepsilon \mapsto \begin{pmatrix} \mathrm{Re}(p_0^2) & -\mathrm{Im}(p_0^2) & \mathrm{Re}(2p_0p_1) \\ \mathrm{Im}(p_0^2) & \mathrm{Re}(p_0^2) & \mathrm{Im}(2p_0p_1) \\ 0 & 0 & 1 \end{pmatrix},$$

where $\mathrm{Re}(2p_0p_1)$ (respectively, $\mathrm{Im}(2p_0p_1)$) is the real (respectively, imaginary) part of $2p_0p_1 \in \mathbb{C}$. Note that this gives a surjective group homomorphism $\varphi : \check{\mathbb{C}}_1 \to \mathrm{SE}(2)$ whose kernel is $\{\pm 1\}$. That is, the preimage of any 2D rigid transformation consists of exactly two unit DCN's with opposite signs.

Example 1 We compute the DCN's $\pm\hat{p} \in \check{\mathbb{C}}_1$ which represent θ-rotation around $v \in \mathbb{C}$. It is the composition of the following three rigid transformations: $(-v)$-translation, θ-rotation around the origin, and v-translation. Thus, we have

$$\hat{p} = \left(1 + \frac{v}{2}\varepsilon\right) \cdot \left(\pm e^{\frac{\theta}{2}i}\right) \cdot \left(1 - \frac{v}{2}\varepsilon\right) = \pm\left(e^{\frac{\theta}{2}i} + \left(e^{-\frac{\theta}{2}i} - e^{\frac{\theta}{2}i}\right)\frac{v}{2}\varepsilon\right).$$

4 Relation to the Dual Quaternion Numbers

The following ring homomorphism

$$\check{\mathbb{C}} \ni p_0 + p_1\varepsilon \mapsto p_0 + p_1 j\varepsilon \in \hat{\mathbb{H}}$$

is compatible with the involution and the conjugation, and preserves the norm. Furthermore, if we identify $v = (x, y) \in \mathbb{R}^2$ with $1 + (xj + yk)\varepsilon$ $(= 1 + (x + yi)j\varepsilon)$, the above map commutes with the action. From this point of view, DCN is nothing but a sub-ring of DQN.

Note also that $\check{\mathbb{C}}$ can be embedded in the ring of the 2×2-complex matrices by

$$p_0 + p_1\varepsilon \mapsto \begin{pmatrix} p_0 & p_1 \\ 0 & \tilde{p}_0 \end{pmatrix}.$$

Then

$$\widetilde{\begin{pmatrix} p_0 & p_1 \\ 0 & \tilde{p}_0 \end{pmatrix}} = \begin{pmatrix} \tilde{p}_0 & p_1 \\ 0 & p_0 \end{pmatrix}, \quad \left|\begin{pmatrix} p_0 & p_1 \\ 0 & \tilde{p}_0 \end{pmatrix}\right|^2 = \det \begin{pmatrix} p_0 & p_1 \\ 0 & \tilde{p}_0 \end{pmatrix}.$$

We thus have various equivalent presentations of DCN. However, our presentation of $\check{\mathbb{C}}$ as the anti-commutative dual numbers is easier to implement and more efficient.

5 Interpolation of 2D Rigid Transformations

First, recall that for the positive real numbers $x, y \in \mathbb{R}_{>0}$, there are two typical interpolation methods:

$$(1 - t)x + ty, \quad t \in \mathbb{R}$$

and

$$(yx^{-1})^t x, \quad t \in \mathbb{R}.$$

The first method can be generalized to DQN as the *Dual quaternion Linear Blending* in [7]. Similarly, a DCN version of *Dual number Linear Blending (DLB, for short)* is given as follows:

Definition 3 For $\hat{p}_1, \hat{p}_2, \ldots, \hat{p}_n \in \check{\mathbb{C}}_1$, we define

$$DLB(\hat{p}_1, \hat{p}_2, \ldots, \hat{p}_n; w_1, w_2, \ldots, w_n) = \frac{w_1\hat{p}_1 + w_1\hat{p}_2 + \cdots + w_n\hat{p}_n}{|w_1\hat{p}_1 + w_1\hat{p}_2 + \cdots + w_n\hat{p}_n|}, \tag{2}$$

where $w_1, w_2, \ldots, w_n \in \mathbb{R}$. Note that the denominator can become 0 and for those set of w_i's and $\hat{p}_i$'s DLB cannot be defined.

A significant feature of DLB is that it is distributive (it is called *bi-invariant* in some literatures). That is, the following holds:

$$\hat{p}_0 DLB(\hat{p}_1, \hat{p}_2, \ldots, \hat{p}_n; w_1, w_2, \ldots, w_n) = DLB(\hat{p}_0\hat{p}_1, \hat{p}_0\hat{p}_2, \ldots, \hat{p}_0\hat{p}_n; w_1, w_2, \ldots, w_n).$$

This property is particularly important when transformations are given in a certain hierarchy such as in the case of skinning; if the transformation assigned to the root joint is modified, the skin associated to lower nodes is deformed consistently.

Next, we consider the interpolation of the second type. For this, we need the exponential and the logarithm maps for DCN.

Definition 4 For $\hat{p} = p_0 + p_1\varepsilon \in \check{\mathbb{C}}$, we define

$$\exp \hat{p} = \sum_{n=0}^{\infty} \frac{(p_0 + p_1\varepsilon)^n}{n!} = e^{p_0} + \frac{(e^{p_0} - e^{\tilde{p}_0})}{p_0 - \tilde{p}_0} p_1\varepsilon.$$

When $\exp \hat{p} \in \check{\mathbb{C}}_1$, we can write $p_0 = \theta i$ for some $-\pi \leq \theta < \pi$, and

$$\exp \hat{p} = e^{\theta i} + \frac{\sin\theta}{\theta} p_1\varepsilon.$$

For $\hat{q} = e^{\theta i} + q_1\varepsilon \in \check{\mathbb{C}}_1$, we define

$$\log(\hat{q}) = \theta i + \frac{\theta}{\sin\theta} q_1\varepsilon.$$

As usual, we have $\exp(\log(\hat{q})) = \hat{q}$ and $\log(\exp(\hat{p})) = \hat{p}$. Note that this gives the following *Lie correspondence* (see [3, 8])

$$\exp : \mathfrak{dcn} \to \check{\mathbb{C}}_1,$$
$$\log : \check{\mathbb{C}}_1 \to \mathfrak{dcn},$$

where $\mathfrak{dcn} = \{\theta i + p_1\varepsilon \in \check{\mathbb{C}} \mid \theta \in \mathbb{R}, p_1 \in \mathbb{C}\} \simeq \mathbb{R}^3$. We have the following commutative diagram:

$$\begin{array}{ccc} \mathfrak{dcn} & \xrightarrow{d\varphi} & \mathfrak{se}(2) \\ \downarrow{\scriptstyle \exp} & & \downarrow{\scriptstyle \exp} \\ \check{\mathbb{C}}_1 & \xrightarrow{\varphi} & SE(2), \end{array}$$

where

$$d\varphi : \mathfrak{dcn} \to \mathfrak{se}(2)$$

Table 1 Comparison of computational cost

	Transformation (FLOPs)	Composition (FLOPs)	Conversion (FLOPs)	Memory usage (scalars)
DCN	22	20	15	4
DQN	92	88	NA	8
2 × 2-complex matrix	112	56	15	8
3 × 3-real matrix	15	45	18 (to DCN)	9

$$\theta i + (x + yi)\varepsilon \mapsto \begin{pmatrix} 0 & -2\theta & 2x \\ 2\theta & 0 & 2y \\ 0 & 0 & 0 \end{pmatrix},$$

is an isomorphism of $\mathbb{R}$-vector spaces.

The following is a DCN version of SLERP [12].

Definition 5 SLERP interpolation from $\hat{p} \in \check{\mathbb{C}}_1$ to $\hat{q} \in \check{\mathbb{C}}_1$ is given by

$$SLERP(\hat{p}, \hat{q}; t) = (\hat{q}\hat{p}^{-1})^t \hat{p} = \exp(t \log(\hat{q}\hat{p}^{-1}))\hat{p},$$

where $t \in \mathbb{R}$.

This gives a uniform angular velocity interpolation of two DCN's, while DLB can blend three or more DCN's without the uniform angular velocity property.

6 Computational Cost

We compare the following four methods for 2D rigid transformation: our DCN, the unit dual quaternion numbers (see Sect. 4), the 3 × 3-real (homogeneous) matrix representation of SE(2), and the 2 × 2-complex matrices described below.

In (Table 1), we list the computational cost in terms of the number of floating point operations for

- transforming a point
- composing two transformations
- converting a transformation to the standard 3 × 3-real matrix representation (except for the 3 × 3-real matrix case where it shows the computational cost to convert to DCN representation).

We also give the memory usage for each presentation in terms of the number of floating point units necessary to store a transformation.

Note that when a particular application requires to apply a single transformation to a lot of points, it is faster to first convert the DCN to a 3 × 3-real matrix.

7 A C++ Library

We implemented our DCN. Though it is written in C++, it should be easy to translate to any language. You can download the MIT-licensed source code at [13]. We also included a small demo application for iOS.

Acknowledgments This work was supported by Core Research for Evolutional Science and Technology (CREST) Program "Mathematics for Computer Graphics" of Japan Science and Technology Agency (JST). The authors are grateful for S. Hirose at OLM Digital Inc., and Y. Mizoguchi, S. Yokoyama, H. Hamada, and K. Matsushita at Kyushu University for their valuable comments.

References

1. Alexa M, Cohen-Or D, Levin D (2000) As-rigid-as-possible shape interpolation. In: Proceedings of the 27th annual conference on computer graphics and interactive techniques (SIGGRAPH '00). ACM Press/Addison-Wesley Publishing Co., New York, NY, USA, pp 157–164. doi:10.1145/344779.344859
2. Gleicher M, Witkin A (1992) Through-the-lens camera control. SIGGRAPH Comput. Graph 26(2):331–340. doi:10.1145/142920.134088
3. Humphreys JE (1978) Introduction to Lie algebras and representation theory. Second printing, revised. Graduate texts in mathematics, vol 9. Springer, New York
4. Igarashi T, Moscovich T, Hughes JF (2005) As-rigid-as-possible shape manipulation. In: Gross M (ed) ACM SIGGRAPH 2005 Papers (SIGGRAPH '05), ACM, New York, NY, USA, pp 1134–1141. doi:10.1145/1186822.1073323
5. Jeffers J (2000) Lost theorems of geometry. Am Math Monthly 107(9):800–812
6. Kaji S, Hirose S, Ochiai H, Anjyo K (2013) A Lie theoretic parameterization of Affine transformation. In: Proceedings of MEIS2013 (Mathematical Progress in Expressive Image, Synthesis 2013), pp 134–140
7. Kavan L, Collins S, Zara J, O'Sullivan C (2008) Geometric skinning with approximate dual quaternion blending. ACM Trans Graph 27(4):105
8. Ochiai H, Anjyo K (2014) Mathematical formulation of motion and deformation and its applications. In: Mathematical Progress in Expressive Image Synthesis I, Springer-Verlag, Berlin, Germany
9. Ochiai H, Anjyo K (2013) Mathematical description of motion and deformation. In: SIGGRAPH Asia 2013 Course, http://portal.acm.org. Accessed 14 Mar 2014 16h30. (Revised course notes are also available at http://mcg.imi.kyushu-u.ac.jp/english/project.php?record_id=95. Accessed 14 Mar 2014 16h30)
10. Pinheiro S, Gomes J, Velho L (1999) Interactive specification of 3D displacement vectors using arcball. In: Proceedings of the international conference on computer graphics (CGI '99). IEEE Computer Society, Washington, DC, USA, p 70
11. Schaefer S, McPhail T, Warren J (2006) Image deformation using moving least squares. In: ACM SIGGRAPH 2006 Papers (SIGGRAPH '06). ACM, New York, NY, USA, pp 533–540. doi:10.1145/1179352.1141920
12. Shoemake K (1985) Animating rotation with quaternion curves. In: ACM SIGGRAPH, pp 245–254
13. https://github.com/KyushuUniversityMathematics/iPad-ProbeDeformer. Accessed 14 Mar 2014 16h30

Phase Dynamics on the Modified Oscillators in Bipedal Locomotion

Wulin Weng, Shin-Ichiro Ei and Kunishige Ohgane

Abstract Based on neurophysiological evidence, studies modeling human locomotion system have shown that a bipedal walking is generated by mutual entrainments between the oscillatory activities of a central pattern generator (CPG) and a Body. The walking model could well reproduce human walking. However, it has been also shown that time delay in the sensorimotor loop destabilizes mutual entrainments, which leads a failure to walk. Recently, theoretical studies have discovered a phenomenon in which a CPG can induce the phase of its oscillatory activity to shift forward according to time delay. This self-organized phenomenon overcoming time delay is called "*flexible-phase locking*". Then, theoretical studies have hypothesized that one of the essential mechanisms to yield of *flexible-phase locking* is a stable limit cycle of CPG activity. This study demonstrates the hypothesis in walking models through computer simulation by replacing the CPG model with the one having different oscillation properties.

Keywords CPG · Body · Time delay · Limit cycle · Flexible-phase locking

---

W. Weng (✉)
Graduate School of Mathematics, Kyushu University, 744, Motooka, Nishi-ku, Fukuoka 819-0395, Japan
e-mail: b-oh@math.kyushu-u.ac.jp

S.-I. Ei
Institute of Mathematics for Industry, Kyushu University, 744, Motooka, Nishi-ku, Fukuoka 819-0395, Japan
e-mail: ichiro@imi.kyushu-u.ac.jp

K. Ohgane
Graduate School of Human Sciences, Kinjo Gakuin University, 2-1723, Omori, Moriyama-ku, Aichi 463-8521, Japan
e-mail: ohganemath@yahoo.co.jp

K. Anjyo (ed.), *Mathematical Progress in Expressive Image Synthesis I*,
Mathematics for Industry 4, DOI: 10.1007/978-4-431-55007-5_18,

1 Introduction

For motor control, time delays in signal transmission through a sensorimotor loop cause a serious problem in general. It is well known that the loop time delay in human locomotion systems is very long [1–4]. How a human overcomes the loop time delay is a significant and important problem.

Based on neurophysiological evidences, studies for models of human locomotion system have demonstrated that a bipedal walking is generated by a mutual entrainment of oscillatory activities of a central pattern generator (CPG) and a musculoskeletal system (Body) [5, 6]. The model could well reproduce human walking. However, the model has also shown [7]) that even a short time delay (70 ms) in a sensorimotor loop between the CPG and Body causes the failure to walk. On the other hand, Ohgane et al. [8] have found that the system structure, i.e., the mutual entrainment between the CPG and Body [5] can overcome loop time delays by an emergent adaptive phenomenon. It was called *flexible-phase locking* in [8], in which the phase of CPG activities shifts forward according to an interval of loop time delay. Emergence of flexible-phase locking has been discovered by the modification of the walking model. The main point of the modification was the replacement of neuron models to compose a CPG. While the previous studies (Taga et al. [5]) has used harmonic oscillators [9, 10] to compose the CPG, Ohgane et al. [8] have modeled the CPG by Bonhöffer van del Pol (BVP) equations [11] which is known as a physiologically faithful neuron model. The mechanisms causing flexible-phase locking are interesting from the viewpoint of motor control or neuroscience. Simplifying the waking model yielding flexible-phase locking [8] and analyzing the simplified model, Ohgane et al. have also argued in [12] that one of the mechanisms for causing flexible-phase locking is a stable limit cycle of CPG activity. However, the simplification has been based on only a few characteristics of walking models and conveniences for analysis. That is, the simplified model has been not described by the reduction of the walking model. Whether the mechanisms understood by the analysis of the simplified model have an universality in human locomotion phenomena or not is an open question. That is, whether the similar mechanisms induce flexible-phase locking in human locomotion model or not is an inevitable problem.

The purpose of this chapter is a direct confirmation of it. We investigate the validity of neuronal limit cycles as a mechanism yielding flexible-phase locking in human walking model. Replacing the description of CPG neurons by ones having different oscillation properties, we can observe phase shift behaviors of the CPGs. Using two kinds of limit cycle oscillators with different properties mentioned below, we confirm the behaviors of phase shift. The one is $\lambda - \mu$ system [13–15], in which the orbit pattern is of just harmonic oscillation but it is originally stable. The other is Van der Pol oscillator [16], in which the limit cycle orbit is formed typically by a distortion of the orbit of harmonic oscillations. Replacing CPG neurons with these oscillators, we also observe the occurrence of flexible-phase locking of the CPG in the walking model with loop time delays.

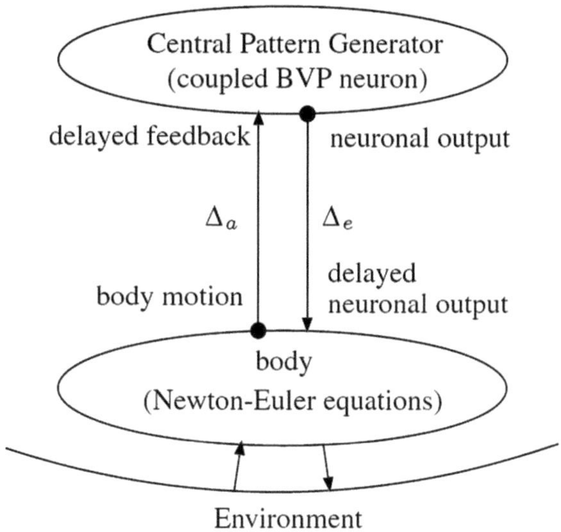

Fig. 1 Outline of the walking model (a delayed synaptically coupling of CPG and Body.) Δ_a is an afferent time delay and Δ_e is an efferent time delay. This figure is from [8]

2 Flexible-Phase Locking in the Walking Model

2.1 Walking Model

In this section, we introduce a walking model in which flexible-phase locking can occur [8]. We called this model *the delayed direct-coupled CPG and Body* (Fig. 1).

The time delays through the sensorimotor loops are assumed to be represented as follows: The total time delay Δ_t through the loop consists of two equivalent amounts of time delay $\Delta_t = \Delta_a + \Delta_e$, i.e., an afferent delay Δ_a and an efferent delay Δ_e [7]. Moreover, $\Delta_a = \Delta_e (=: \Delta)$ was assumed for simplicity [17, 18].

The equations for Body are represented by means of the Newton-Euler method (Appendix). We denote $\mathbf{x} = (x_1, \ldots, x_6) \in \mathbf{R}^6$ a vector of the mass point positions of 1 link and the inertial angles of 4 links as shown in Fig. 9.

The CPG composed of 12 neurons (Fig. 2) is represented by the following coupled BVP (Bonhöffer van del Pol) differential equations [11]:

$$\begin{cases} \tau_i \dot{u_i}(t) = u_i(t) - v_i(t) - u_i(t)^3/3 + \sum_{ij=1}^{12} (w_{ij} y_i) + u_0 + \alpha_w F_i(\mathbf{X}(t - \Delta)), \\ \tau_i' \dot{v_i}(t) = u_i(t) + a - b v_i(t), \\ y_i = f(u_i(t)), \ f(u) = \max(0, u) \ (i = 1, 12), \end{cases} \tag{1}$$

where u_i is the potential of the ith neuron; v_i is responsible for the accommodation and refractoriness of the i-th neuron; w_{ij} denote connecting weight from the i-th neuron to the j-th neuron; τ_i and τ_i' are the time constants of the potential and

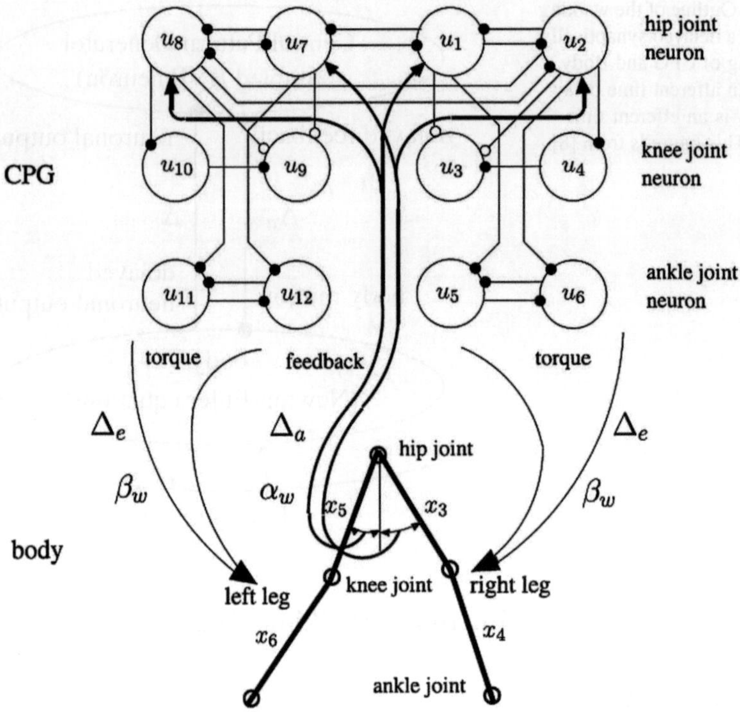

Fig. 2 Ohgane's walking model [8] which can yield flexible-phase locking. The central pattern generator (CPG) has been described by using Bonhöffer van del Pol (BVP) equation which is known as physiologically faithful neuron model. The CPG, Body, and feedback pathway are shown. u_i is the potential of the i-th neuron in the CPG. ∘ and • denote an excitatory connection and an inhibitory connection, respectively. x_3, x_4, x_5, and x_6 are the angles of Body segments. The motion of the hip, the *knee joint* and the *ankle joint* in the *right leg* are governed by neurons 1–2, 3–4, and 5–6, respectively. Similarly, the motion of the joints in the *left leg* is governed by neurons 7–12. Odd-numbered neurons control flexions of the joint, while even-numbered neurons control its extension. The *hip joint* angles of both legs are used as feedback to the hip joint neurons. The afferent delay Δ_a and the efferent delay Δ_e take place in the transmission of the neuronal output and of the feedback, respectively. α_w and β_w are the afferent coupling strength and the efferent coupling strength. It is confirmed by computer simulations that the CPG itself has an asymptotically stable limit cycle. This figure is from [8]

the accommodation and refractory effects, respectively; y_i is the output of the i-th neuron; u_0 is a constant parameter. Δ is a time delay. α_w is a positive coefficient of afferent connections from the Body to the CPG; F_i is a sensory feedback; t is time; a and b are positive constants; the natural frequency of each joint neuron (τ_i, τ_i') is a set of values similar to the natural frequency of each joint angle [5].

The CPG and Body interact by a torque from the Body (T_r in Appendix) and sensory feedbacks $F_i (i = 1, \ldots, 12)$. The torque and sensory feedbacks reach with time delays (Δ_a and Δ_e in Fig. 1). The torque T_r was assumed to be proportional to the magnitude of the neuronal output from the CPG.

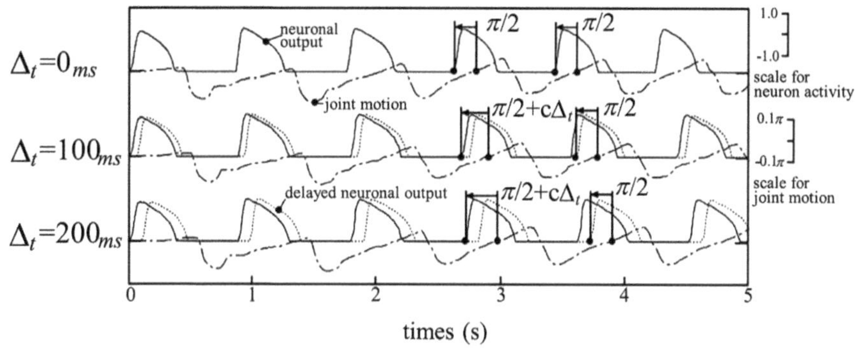

Fig. 3 Flexible-phase locking of the CPG neuron described by BVP equations in the walking model [8] shown in Fig. 2. The graph shows the activities of flexor neuron (y_1) and angle motion (x_5) in the left leg. *Solid line*, *dotted line*, and *dot-dashed line* denote the neuronal output, the delayed neuronal output, and the joint angle motion, respectively. *Arrows* indicate the phase difference between neuronal output and joint angle motion, and between delayed neuronal output and joint angle motion. The phase of neuronal output is shifted forward according to Δ_t; t_1, t_2 are the times represented in this figure, and c is a constant value. Therefore, the phase relationship between delayed neuronal output and joint angle motion is maintained constant. This figure is from [12]

The CPG receives sensory feedback from the Body. The feedback $F_i (i = 1, \ldots, 12)$ to the i-th neuron is given as follows:

$$F_1 = F, F_2 = F', F_7 = F', F_8 = F, \quad F_i = 0 \ \text{(else)},$$

where F, F' are given as follows:

$$F = -f(-x_5(t - \Delta)), F' = -f(-x_3(t - \Delta)), f(x) = \max(0, x),$$

where x_3 and x_5 are the thigh angles of the right and left legs at the hip joint, respectively, as defined in Appendix.

The walking model has shown behaviors adaptive to time delay. The system successfully walks by a forward phase shift of the CPG activity, according to the time delay interval as shown in Figs. 3 and 4. The results (Figs. 3, 4) are simulated by C programming and Newton-Euler method (see Appendix).

We call such adaptive phase shift *flexible-phase locking*. Investigating mechanisms of flexible-phase locking, the previous theoretical studies [12] have argued that one of the essential mechanisms is a stable limit cycle of CPG activity. If the mechanism is true for the flexible-phase locking, similar structures can be observed even if the limit cycle of the CPG is replaced with other limit cycles. In the next section, we demonstrate it by replacing the oscillator of the CPG with other two oscillators.

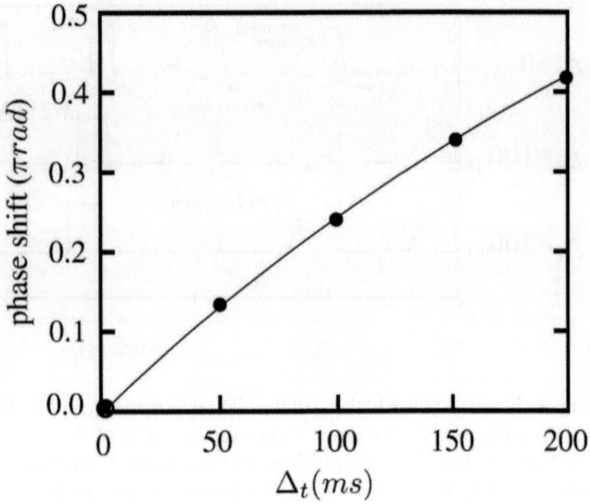

Fig. 4 The forward phase shift of neuronal output of the CPG described by BVP equations in the walking model [8] is shown as a function of the total time delay Δ_t. The vertical axis denotes $\frac{t_2 - t_1}{T} - \frac{\pi}{2}$. T is a period of walking cycle. t_1, t_2 are the times represented in Fig. 3. The phase of neuronal output is shifted forward according to Δ_t. This figure is from [12]

3 The Walking Systems with Other CPG Models

3.1 $\lambda - \mu$ System

The $\lambda - \mu$ system is

$$\begin{cases} \dot{u} = \varepsilon(\lambda - u^2 - v^2)u - \mu v, \\ \dot{v} = \varepsilon(\lambda - u^2 - v^2)v + \mu u \end{cases} \tag{2}$$

for positive constants $\varepsilon > 0$, $\lambda > 0$ and $\mu > 0$. It is easy to see that (2) has a stable periodic solution $S(t) = \sqrt{\lambda}(\cos \mu t, \sin \mu t)$ with the period $p := 2\pi/\mu$ and amplitude $\sqrt{\lambda}$.

In this subsection, we revise Ohgane's model (1) by replacing BVP equations with $\lambda - \mu$ system. Through computer simulations, we confirm that the revised model could also overcome loop time delays by flexible-phase locking. The revised CPG model is represented by the following differential equations:

$$\begin{cases} \dot{u_i}(t) = \varepsilon_i(\lambda_i - u_i^2(t) - v_i^2(t))u_i(t) - \mu_i v_i(t) + u_0 + \alpha_w F_i(\mathbf{x}(t - \Delta)), \\ \dot{v_i}(t) = \varepsilon_i(\lambda_i - u_i^2(t) - v_i^2(t))v_i(t) + \mu_i u_i(t), \end{cases} \tag{3}$$

where ε_i, λ_i and μ_i are the positive parameters.

Here we denote that in no time delay condition $\Delta_t = 0ms$, the system walks stably.

In various conditions of time delay intervals, the system could walk by shifting forward of its CPG activity, as shown in Figs. 5 and 6.

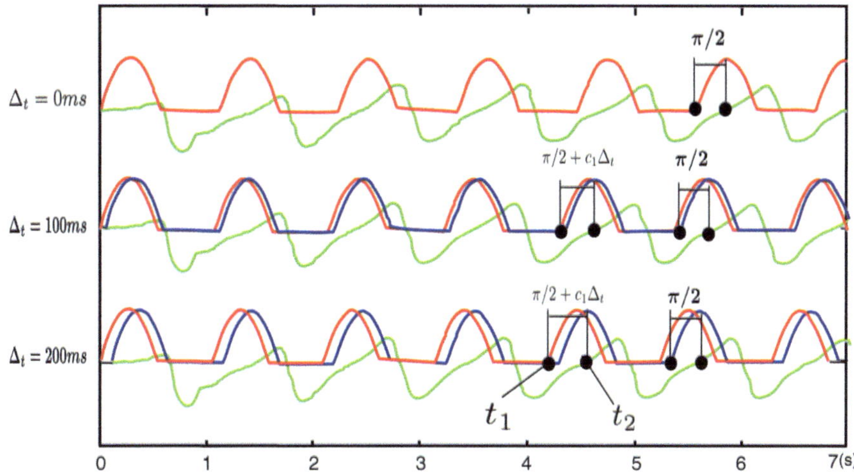

Fig. 5 Flexible-phase locking caused by the CPG composed of $\lambda - \mu$ systems (3) in the walking model. The graph shows the activities of flexor neuron (y_1) and angle motion (x_5) in the left leg. *Red line*, *green line*, and *blue line* denote the neuronal output, the joint angle motion, and the delayed neuronal output, respectively. where the parameters $\varepsilon_i = 1.0$, $\lambda_i = 1.0$, $\mu_i = 5.0$. The phase of neuronal output is shifted forward according to Δ_t. t_1, t_2 are the times represented in this figure and c_1 is a constant value. Therefore, the phase relationship between delayed neuronal output and joint angle motion is maintained constant

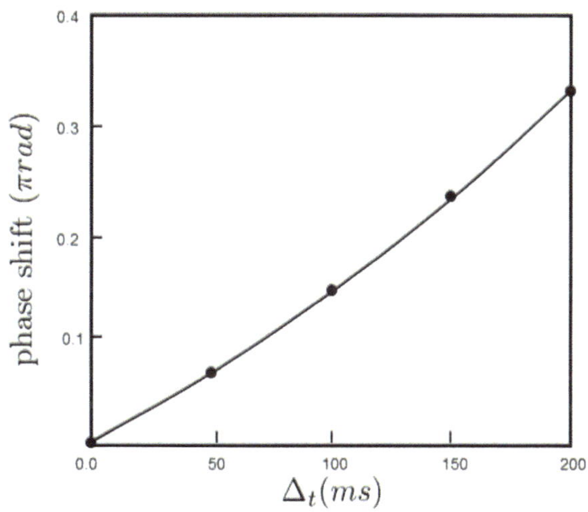

Fig. 6 The forward phase shift of neuronal output caused by the CPG composed of $\lambda - \mu$ systems (3) in the walking model are shown as a function of the total time delay Δ_t. The *vertical axis* denotes $\frac{t_2 - t_1}{T} - \frac{\pi}{2}$. T is a period of walking cycle. t_1, t_2 are the times represented in Fig. 5

3.2 Van der Pol Oscillator

The Van der Pol oscillator also has a stable limit cycle. Its stability of the orbit is produced by nonlinear damping governed by the second-order differential equation.

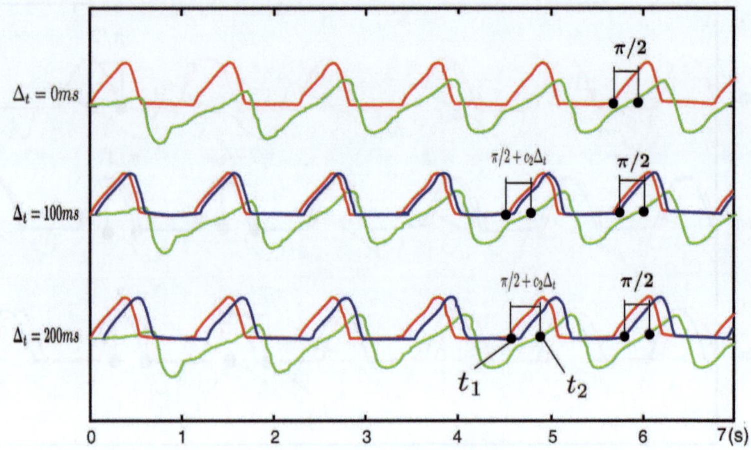

Fig. 7 Flexible-phase locking caused by the CPG composed of Van der Pol oscillators (5) in the walking model. The graph shows the activities of flexor neuron (y_1) and angle motion (x_5) in the left leg. *Green line*, *red line*, and *blue line* denote the neuronal output, the joint angle motion, and the delayed neuronal output, respectively. The phase of neuronal output is shifted forward according to Δ_t. $\varepsilon_i = 1.0$, t_1, t_2 are the times represented in this figure. c_2 represents a constant value. Therefore, the phase relationship between delayed neuronal output and joint angle motion is maintained constant

$$\ddot{u} - \varepsilon(1 - u^2)\dot{u} + u = 0,$$

where u is the dynamical variable and $\varepsilon > 0$ a parameter.

Another commonly used form based on the transformation $v = \dot{u}$ is leading to

$$\begin{cases} \dot{u} = v, \\ \dot{v} = \varepsilon(1 - u^2)v - u. \end{cases} \tag{4}$$

In this subsection, we revise Ohgane's model (1) by replacing BVP equations (1) with Van der Pol oscillator. Through computer simulations, we confirm that the revised model can also overcome loop time delays by flexible-phase locking. The revised CPG model is represented by the following differential equations:

$$\begin{cases} \dot{u_i}(t) = v_i(t) + \sum_{ij=1}^{12}(w_{ij}y_i) + u_0 + \alpha_w F_i(\mathbf{x}(t - \Delta)), \\ \dot{v_i}(t) = \varepsilon_i(1 - u_i^2(t))v_i(t) - u_i(t), \end{cases} \tag{5}$$

where ε is the positive parameter.

Here we denote that in no time delay condition $\Delta_t = 0\,\text{ms}$, the system walks stably.

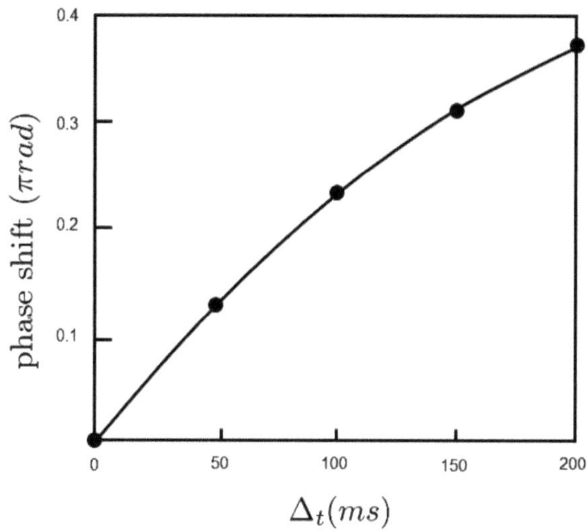

Fig. 8 The forward phase shift of neuronal output caused by the CPG composed of Van der Pol oscillators (5) in the walking model are shown as a function of the total time delay Δ_t. The vertical axis denotes $\frac{t_2 - t_1}{T} - \frac{\pi}{2}$. T is a period of walking cycle. t_1, t_2 are the times represented in Fig. 7

In various conditions of time delay intervals, the system could walk by shifting forward of CPG activity, as shown in Figs. 7 and 8.

4 Conclusions

As conjectured in the previous studies, one of the essential mechanisms causing flexible-phase locking is a stable limit cycle of CPG activity. In this paper, in order to confirm it, we revised Ohgane's walking model [8] by replacing the CPG model. For the replacement of the CPG model, we prepared two other CPG models that have different machinery for production of limit cycle each other. Moreover, they can nearly harmonically oscillate by modulation of their parameters. Computer simulations of the two revised walking models suggested that the walking system can yield flexible-phase locking when the CPG oscillates as stable limit cycles, regardless of the mechanism of limit cycle; flexible-phase locking becomes hard to occur when the stability of limit cycle of CPG oscillation weakens. Thus, we confirmed the validity of the conjecture that one of the essential mechanism for the yield of flexible-phase locking is a stable limit cycle of CPG activity. This mechanism is suggested to have an universality in bipedal walking models. Further investigations on the mechanism causing flexible-phase locking are in progress.

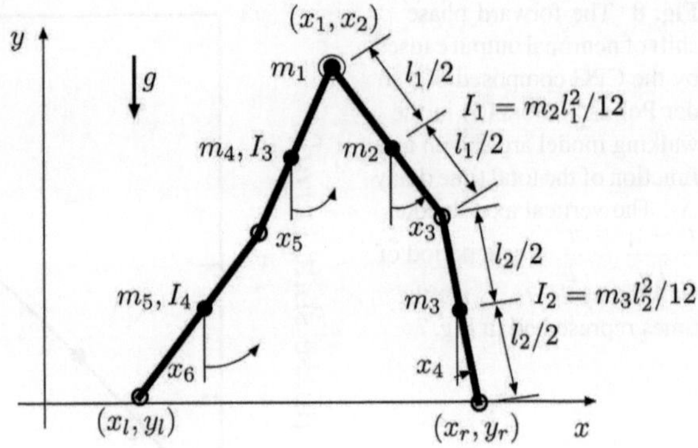

Fig. 9 Model of bipedal Body as an interconnected chain of five rigid links (a point mass m_1 on the hip and four rigid bodies $I_i (i = 1, 4)$

Appendix: The Body of the Walking Model

All variables and conventions correspond to those shown in Fig. 9. All variables and parameters follow the ones proposed by Taga (1994). By using the Newton-Euler method, motion of the Body (Taga, 1991) can be written as follows:

$$P(\mathbf{x})\ddot{\mathbf{x}} = \mathbf{Q}(\mathbf{x}, \ddot{\mathbf{x}}, T_r(\mathbf{y})),$$

therefore,

$$\ddot{\mathbf{x}} = [P(\mathbf{x})]^{-1}\mathbf{Q}(\mathbf{x}, \ddot{\mathbf{x}}, T_r(\mathbf{y})).$$

References

1. Hammond PH (1956) The influence of prior instruction to the subject on an apparently involuntary neuro-muscular response. J Physiol 132:17–18
2. Chan CWY, Melvill JG, Kearney RE, Watt DG et al (1979) The late electromyographic response to limb displacement in man. I. Evidence for supraspinal contribution. Electroencephalogr Clin Neurophysiol 46:173–181
3. Chan CWY, Melvill JG, Kearney RE, Watt DG et al (1979) The late electromyographic response to limb displacement in man. ii. sensory origin. Electroencephalogr Clin Neurophysiol 46:182–188
4. Shinoda Y, Yamaguchi T, Futami T et al (1986) Multiple axon collaterals of single corticospinal axons in the cat spinal cord. J Neorophysiol 55:425–448
5. Taga G, Yamaguchi Y, Shimizu H et al (1991) Self-organized control of bipedal locomotion by neural oscillators in unpredictable environment. Biol Cybern 65:147–159
6. Grillner S (1985) Neurobiological bases of rhythmic motor acts in vertebrates. Science 228:143–149
7. Taga G (1994) Emergence of bipedal locomotion through entrainment among the neuro-musculo-skeltal system and environment. Phys D 75:190–208
8. Ohgane K, Ei SI, Kudo K, Ohtsuki T et al (2004) Emergence of adaptability to time delay in bipedal locomotion. Biol Cybern 90:125–132

9. Matsuoka K (1985) Sustained oscillations generated by mutually inhibiting neurons with adaptation. Biol Cybern 52:367–376
10. Matsuoka K (1987) Mechanisms of frequency and pattern control in the neural rhythm generators. Biol Cybern 56:345–353
11. Fitzhugh R (1961) Impulses and physiological states in theoretical models of nerve membrane. Biophys 1:445–466
12. Ohgane K, Ei SI, Mahara H (2009) Neuron phase shift adaptive to time delay in locomotor control. Appl Math Model 33:797–811
13. Ei SI (2004) A remark on the interpretation of periodic solutions. Bull Jpn Soc Ind Appl Math 14(1):35–47
14. Ei SI, Ohgane K (2011) A new treatment for periodic solutions and coupled oscillators. Kyushu J Math 65:197–217
15. Weng WW, Ei SI, Ohgane K (2012) The functional roles of time delay on flexible phase-locking in bipedal locomotion. J Math-for-Industry 4:123–133
16. Van der Pol B (1920) A theory of the amplitude of free and forced triode vibrations. Radio Rev 1(701–710):754–762
17. Caruso G, Labianca O, Ferrannini E et al (1973) Effect of ischemia on sensory potentials of normal subjects of different ages. J Neurol Neurosurg Psychiatry 36:455–466
18. Mankovskij NB, Timko NA et al (1973) Age-related characteristics of the functional condition of the neoromuscular system. Z Alterforsch 27:191–200

Part V
Image Database and Applications

Single-View 3D Reconstruction by Learning 3D Game Scenes

Makoto Okabe, Ken Anjyo and Rikio Onai

Abstract We report a method for reconstructing a three-dimensional (3D) depth map using a single two-dimensional (2D) image. Our method is designed to reconstruct manmade objects, such as buildings. We first estimate the normal map, and then integrate it to obtain the depth map. To estimate the normal map, we analyze the co-occurrence relation between the normal vectors and image features in a training dataset. We consider the corners and lines to be the image features. The training dataset is formed of 3D game scenes. In the offline learning phase, we detect corners and lines in the normal map of each game scene using a detection algorithm, and observe the normal vectors around them. Then we construct a database of the co-occurrence relations, i.e., how frequently each corner or line appears with each normal vector. In the online reconstruction phase, given an input image, we detect the corners and lines using the same detection algorithm, and estimate the normal vectors around them based on the learned co-occurrence relation. We formulate this estimation using a Markov random field. Finally, the estimated normal map is integrated by solving Poisson's equation, and we obtain a depth map.

Keywords Single-view modeling · Tour into the picture · Image-based modeling · Image database · Markov random field · Corner detection

M. Okabe (✉)
The University of Electro-Communications/JST CREST, 2-11, Fujimicho,
Chofu, Tokyo 182-0033, Japan
e-mail: m.o@acm.org

K. Anjyo
OLM Digital, Inc./JST CREST, Dens Kono Bldg. 302, 1-8-8, Wakabayashi,
Setagaya-ku, Tokyo 154-0023, Japan
e-mail: anjyo@olm.co.jp

R. Onai
The University of Electro-Communications, 2-11, Fujimicho, Chofu, Tokyo 182-0033, Japan
e-mail: onai@cs.uec.ac.jp

K. Anjyo (ed.), *Mathematical Progress in Expressive Image Synthesis I*,
Mathematics for Industry 4, DOI: 10.1007/978-4-431-55007-5_19,

1 Introduction

Three-dimensional (3D) geometry is one of the most important cues for scene understanding, and thus 3D reconstruction has been one of the most actively researched areas in the fields of computer vision and graphics for decades. Multi-view stereo techniques, such as those used by Google Maps—Photo Tours [1], and depth sensors, such as Microsoft's Kinect, have become powerful tools for precisely reconstructing scene geometries. However, single-view reconstruction techniques are also popular and have been the subject of much recent research. It is not straightforward to reconstruct precise scene geometries; however, because only a single image is required, it is always easy and intuitive to use. No special hardware is required, and so the technique is generally inexpensive to implement. Animations in which the viewpoint of a single image is changed always have visual impact [2–6].

2 System Overview

First we analyze the co-occurrence relation between the normal vectors and the image features in the training dataset. Our training dataset was a set of normal maps of game scenes, as shown in Fig. 1a. In Fig. 1, the normal map is color-coded, i.e., each normal vector, (n_x, n_y, n_z), at a pixel becomes proportional to the magnitude of the RGB color vector (r, g, b). In the offline-learning phase, we detect corners (see Fig. 1b) and lines (see Fig. 1c) in the normal map of each game scene using an corner-detection algorithm. Then we observe the normal vectors around them, and construct a database of the co-occurrence relation, i.e., how frequently each corner or line appears with each normal vector (see Fig. 1d, e).

In the online reconstruction phase, given the input image shown in Fig. 1f, we can detect corners and lines using the same detector (see Fig. 1g, h). Then we estimate the normal vectors around them based on the learned co-occurrence relation (see Fig. 1i). We consider this estimation an assignment problem of normal vectors to pixels of the input image. We formulate this assignment problem using the Markov random field (MRF), where corners function as data terms, and lines function as smoothness terms. Finally, we integrate the estimated normal map by solving Poisson's equation, and obtain the depth map shown in Fig. 1j.

3 Training Data

The training dataset was a set of normal maps of game scenes (see Fig. 1a). A sample scene in our training dataset is shown in Fig. 2a, and the corresponding depth and normal maps are shown in Fig 2b, c. We used the 3D city-building game Anno 1404, published by UBISOFT [7], to gather our training data. We captured 236 scenes in the

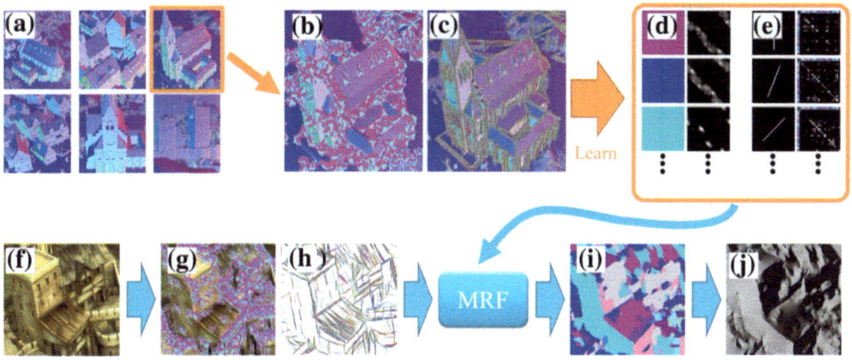

Fig. 1 Overview of the offline-learning phase (*top row*) and the online reconstruction phase (*bottom row*). The normal map is color-coded, i.e., each normal vector, (n_x, n_y, n_z), at a pixel becomes proportional to the magnitude of the RGB color vector (r, g, b)

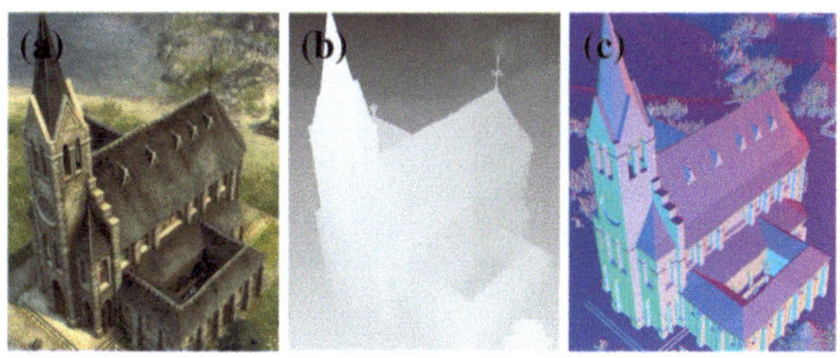

Fig. 2 A sample scene in our training dataset

3D game world. Because we want to reconstruct manmade objects, we intentionally chose scenes of multiple buildings.

4 Offline Learning

Our representation of a corner is shown in Fig. 3. Given an image of a Y-junction, such as that shown in Fig. 3a, we can see the three corners of red, yellow, and blue around it, as shown in Fig. 3b. A single corner has two arms, i.e., it can be represented by a pair of angles, θ and ϕ, as shown in Fig. 3c.

4.1 Co-occurrence of Normal Vectors and Corners

We analyzed the co-occurrence relation between normal vectors and corners. We applied the corner-detection algorithm to each normal map in the training dataset. At each detected corner, we observed the normal vectors at the pixels in the sector

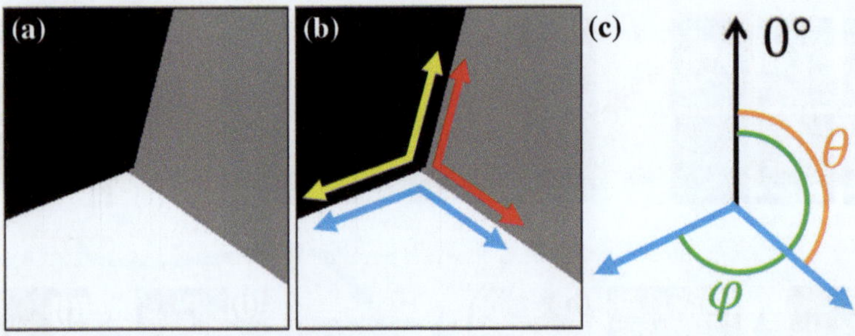

Fig. 3 Our representation of a corner

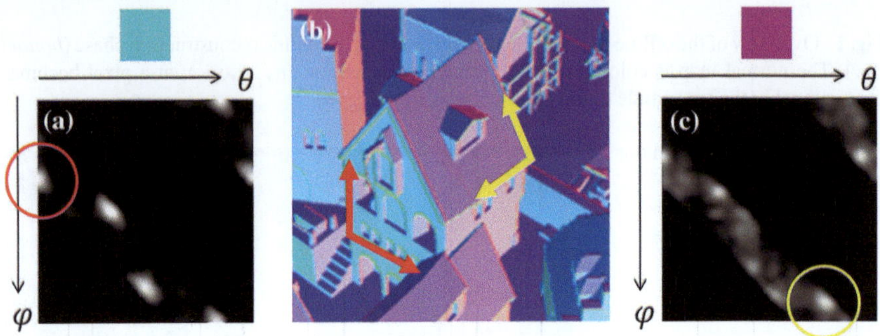

Fig. 4 **a** Map of the frequency of appearance of corners that have *green* normal vectors. The *horizontal* and *vertical* axes represent the direction of the arms, θ and ϕ. **b** The *red* and *yellow* corners correspond to the center pixels of the *red* and *yellow circles* in the maps of the frequency of appearance of corners. **c** Map of the frequency of appearance of corners that have *pink* normal vectors

between the two arms. Figure 4a shows a map of the frequency of appearance of corners that have the green normal vector between their two arms. The horizonal and vertical axes represent the direction of the arms, θ and ϕ, and thus each pixel corresponds to a corner. The pixel intensity is proportional to the frequency of appearance of the corresponding corner.

4.2 Co-occurrence of Normal Vectors and Lines

The line-detection algorithm is similar to that for corner detection. We detect a line by finding a corner where $|\theta - \phi| = 180$. At each detected line, we observe the pair of normal vectors appearing across the line. Figure 5a shows a map of the frequency of appearance of pairs of normal vectors appearing across the vertical line. The horizontal axis represents the index of the normal vectors that appear at the left side of the line. The vertical axis represents the index of the normal vectors that appear at the right side. Each pixel corresponds to a pair of normal vectors appearing at the

Fig. 5 **a** Map of the frequency of appearance of pairs of normal vectors appearing at the *left* and *right* sides of the vertical line. **b** The pair of normal vectors observed across the *red*, *yellow*, and *orange lines* corresponds to the pixel at which the *arrow* of the corresponding color in the maps points. **c** Map of the frequency of appearance of pairs of normal vectors appearing at the *left* and *right* sides of the diagonal line

left and right sides of the line. The pixel intensity is proportional to the frequency of appearance of the corresponding pair of normal vectors.

5 Online Reconstruction

Given a target image, we estimate the normal map assuming that the normal vectors will appear with corners and lines detected in the target image according to the frequency of appearance analyzed as described in the previous section. We formulate the estimation as the assignment problem of an adequate normal vector to each pixel of the target image.

To simplify the assignment problem, we do not consider all of the normal vectors, rather we chose the representative normal vectors. The estimation of the normal map has become simpler. Given a target image with M pixels, the objective of our algorithm is to perform the assignment $A = [n_1 ... n_M]$, i.e., one of the representative normal vectors to each pixel. We wish to estimate the assignment,

$$A^* = argmax_A \prod_p P(n_p|c_p) \left(\prod_{p,q} P(n_p, n_q|l_{pq}) \right)^\lambda, \quad (1)$$

where n_p and n_q are the normal vectors assigned to the neighboring pixels p and q. Equation (1) is the standard formulation of MRF, where $P(n_p|c_p)$ is the data term and $P(n_p, n_q|l_{pq})$ is the smoothness term.

The term $P(n_p|c_p)$ represents the probability that n_p will be assigned to the pixel p subject to the condition that the corner c_p is observed in the pixel p. Applying Bayes' theorem, we factor the term as,

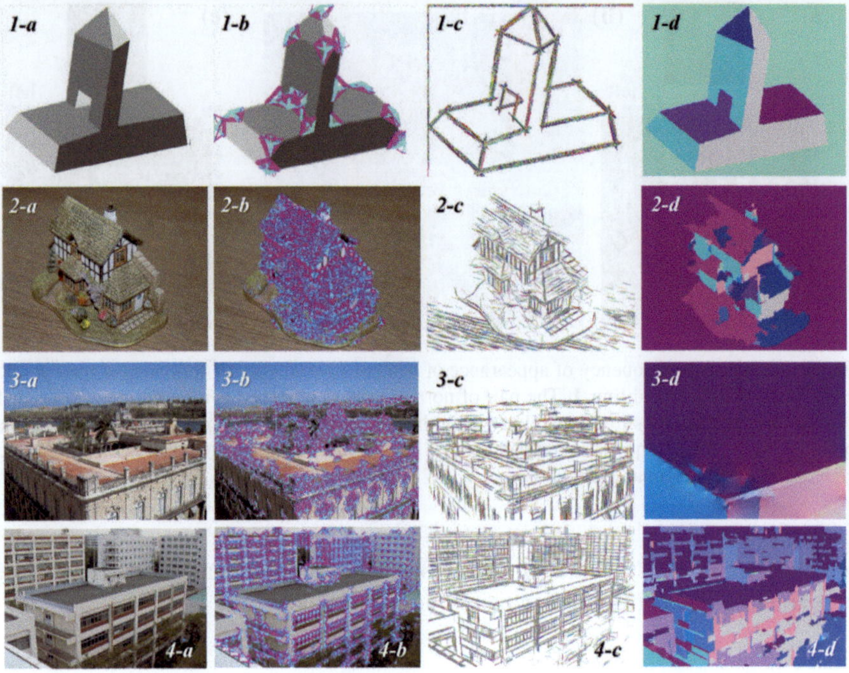

Fig. 6 **a** The input image, **b** detected corners, **c** detected lines, **d** estimated normal map

$$P(n_p|c_p) \propto P(c_p|n_p)P(n_p), \tag{2}$$

where $P(c_p|n_p)$ is the probability distribution of the corner c_p subject to the condition that the normal vector n_p is observed in the pixel p. This probability distribution is equal to the normalized version of the map of the frequency of appearance.

The term $P(n_p, n_q|l_{pq})$ represents the probability that the pair of normal vectors n_p and n_q will be assigned to the neighboring pixels p and q subject to the condition that the line l_{pq} is observed around the pixels p and q. $P(n_p, n_q|l_{pq})$ is equal to the normalized version of the map of the frequency of appearance.

We transform the maximization problem in Eq. (1) into a minimization problem as follows:

$$A^* = argmin_A \sum_p V(n_p) + \lambda \sum_{p,q} W(n_p, n_q), \tag{3}$$

where the data term is $V(n_p) = -log(P(n_p|c_p))$ and the smoothness term is $W(n_p, n_q) = -log(P(n_p, n_q|l_{pq}))$. We solve this energy-minimization problem using the graph-cut method [8], together with the $\alpha - \beta$ swap algorithm.

6 Results and Discussion

Figures 1 and 6 shows the target images, detected corners and lines, and estimated normal maps. The target image of Fig. 6-1-a shows a simple object consisting of a small number of flat surfaces. For this image, our method successfully estimate the reasonable normal map. The estimated normal map consists of three major colors of green, purple, and gray, whose directions are left, upward, and bottom-right. On the other hand, the other target images are photographs. For these images, the estimated normal maps are not perfect but have wrong normal vectors and look noisey. These errors are caused by the errors of detected corners and lines. Since the target image of Fig. 6-1-a is simple and has no detailed texture, the corners and lines are detected almost perfectly shown as Fig. 6-1-b. On the other hand, some of the corners and lines are detected wrongly in the other photographs because of their high frequency textures. One important furture work is to improve the accuracy of the corner and line detection.

References

1. Shan Q, Adams R, Curless B, Furukawa Y, Seitz SM (2013) The visual turing test for scene reconstruction. In: Proceedings of 3DV13
2. Debevec PE, Taylor CJ, Malik J (1996) Modeling and rendering architecture from photographs: a hybrid geometry- and image-based approach. Comput Graph, Ann Conf Ser 30:11–20
3. Horry Y, Anjyo K, Arai K (1997) Tour into the picture: using a spidery mesh interface to make animation from a single image. In: Proceedings of SIGGRAPH '97, pp 225–232
4. Li Z, Guillaume D-P, Jean-Sebastien S, SM Seitz (2001) Single view modeling of free-form scenes. In: Proceedings of CVPR 2001, pp 990–997
5. Hoiem D, Efros AA, Hebert M (2005) Automatic photo pop-up. ACM Trans. Graph. 24(3):577–584
6. Saxena A, Chung SH, Ng AY (2008) 3-D depth reconstruction from a single still image. Int. J. Comput. Vis. 76(1):53–69
7. UBISOFT: Anno 1404 (2010)
8. Szeliski R, Zabih R, Scharstein D, Veksler O, Kolmogorov V, Agarwala A, Tappen M, Rother C (2008) A comparative study of energy minimization methods for Markov random fields with smoothness-based priors. IEEE Trans. Pattern Anal. Mach. Intell. 30(6):1068–1080

Facial Aging Simulation by Patch-Based Texture Synthesis with Statistical Wrinkle Aging Pattern Model

Akinobu Maejima, Ai Mizokawa, Daiki Kuwahara and Shigeo Morishima

Abstract We propose a method for synthesizing a photorealistic human aged-face image based on the patch-based texture synthesis using a set of human face images of a target age. The advantage of our method is that it synthesizes an aged-face image with fine skin texture such as spots and pigments of facial skin, as well as age-related facial wrinkles without blurs (such as those resulting from lack of accurate pixel-wise alignments as in the linear combination model) while maintaining the quality of the original image.

Keywords Automatic · Facial aging · Image patch · Texture synthesis · Wrinkle · Agind pattern · Statistical model

1 Introduction

Predicting the current facial appearance of wanted or missing people who have been missing for several years is an important task in criminal investigation. Previously, sketches or photomontages have been used for this purpose. However, it is not easy to predict and depict one's realistic facial appearance only from photographs that are several years old and some interviews, because the quality of the resulting sketch or montage highly depends on the skill of the forensic artists. Thus, an automatic facial aging simulator, which can synthesize a photorealistic facial appearance based on statistics from the actual aging process, is required.

A. Maejima (✉) · A. Mizokawa · D. Kuwahara
Waseda University, 3-4-1, Okubo, Shinjuku-ku, Tokyo 169-8555, Japan
e-mail: akinobu@mlab.phys.waseda.ac.jp

S. Morishima
Waseda Research Institute for Science and Engineering, Waseda University, 3-4-1, Okubo, Shinjuku-ku, Tokyo 169-8555, Japan
e-mail: shigeo@waseda.jp

K. Anjyo (ed.), *Mathematical Progress in Expressive Image Synthesis I*, Mathematics for Industry 4, DOI: 10.1007/978-4-431-55007-5_20,

Automatic facial aging simulation methods have been proposed by many researchers [2, 3, 5, 6, 9, 10, 12]. Most of the conventional methods adopts a linear combination model such as active appearance model [2, 3, 5, 6, 9] or 3D morphable model [10] to parameterize both facial geometry and appearance. However, in the case of linear combination, which relies on bases, the resulting image is somewhat blurred because the alignment between features, which appear for different individuals or ages is impossible. Burt and Perett reported that perceived age of transformed faces by incorporating the difference between composite faces from different age groups onto real face images. Moreover, they described that the averaging operation involved in creating composite faces was observed to smoothen facial creases and result in composite faces appearing younger than those from their own age groups. Therefore, it is important to represent fine age-related features such as facial wrinkles, pores, pigments which are strongly affected by age perception [1].

To solve this problem, Tazoe et al. have proposed patch-based facial aging texture synthesis using human face images of the target age [12]. This method is based on the assumption that if we accurately reconstruct an input face using image patches of the target age, the resulting face would be the same individual's aged-face. Actually, their method can generate a photorealistic face image with fine age-related features. However, the resulting face sometimes does not look like the original person because the original face has been completely replaced by the reconstructed face. Moreover, age-related features such as wrinkles may be incorrectly expressed depending on the pixel intensity pattern of a patch in the input image.

In this chaper, we propose an automatic facial aging simulation method, which can overcome the above mentioned problems. Our method is based on patch-based texture synthesis using facial images of the target age. Using the statistical wrinkle aging pattern model, we predict the resulting facial wrinkle appearance (shape, depth, and length) of the target age. Moreover, we introduce a modified Poisson solver to seamlessly merge between image patches, and to keep the original facial appearance in the region which influences the individual perception and skin tone.

The principal contribution of this paper is to provide a simulator, which can synthesize photorealistic aged-face images with detailed textures such as spots and pigments of facial skin, as well as age-related facial wrinkles, while maintaining the identity of the original face.

2 Overview

Our facial aging simulation pipeline is depicted in Fig. 1. The facial aging simulation begins with the pre-processing to create patch libraries for age groups and to train the Statistical Wrinkle Aging Pattern Model. Then given an input face image, the actual simulation is done by the runtime processing.

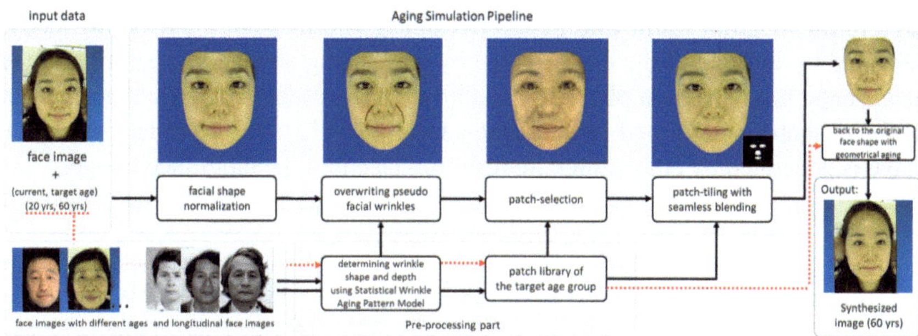

Fig. 1 Our facial aging simulation pipeline. Note that, any images of the target person are not contained in our database for creating path libraries and training the Statistical Wrinkle Aging Pattern Model

Preprocessing:
We collect face images corresponding to various ages, and create an average shaped face model for all faces. Then, shape normalized face image for each face image is generated by the same procedure as the facial shape normalization phase in the runtime processing. Then, we divide shape normalized face images into small square patches with some overlap between neighbors (in this paper, above and to the left). In this paper, the size of each patch is 42 by 42 pixels for 300 by 300 pixel face image and Each patch overlaps each other by 10 pixels. At the same time, an index representing the original position before cutting out are associated with each patch. Then, we construct patch library of age groups for which patches are grouped according to the labeled index. Also, the statistical wrinkle aging pattern model is trained by using longitudinal facial images (details are described in Sect. 4).

Runtime processing:
Shape normalization To remove geometric difference between individuals and to concentrate texture variations, we need to normalize an input facial shape to the average shape. First, 89 facial feature points including eyes, eye brows, nose, lips, and facial contour are detected from the input image using Zhang's technique [13]. Second, the average face model is deformed using Radial Basis Functions [4] so that the vertices of the face model are matched to corresponding detected feature points. Finally, the deformed face model with the input image is rendered into the image. As a result, we can obtain a face image that the shape is normalized to that of the average face.
Overwriting pseudo facial wrinkles Pseudo facial wrinkles generated by the statistical wrinkle aging pattern model (details are described in Sect. 4) consisted of curves representing wrinkles from individual face images over ages are overwritten onto the shape normalized image in order to simulate individual aging, especially with facial wrinkle, which significantly influence human perception of age.
Patch selection and tiling The shape normalized face image is reconstructed by patch-based texture synthesis using patches of the target age (details are described in Sect. 3).

Back to the original shape and geometric aging The shape of the resulting face is deformed toward that of the original face by performing an inverse operation of the facial shape normalization phase. At this time, to represent geometric aging effect, the facial shape is deformed using Tazoe's technique [12]. The complete aged-face image is generated by embedding the resulting face into the input face image.

3 Patch-Based Texture Synthesis Using Age-Specific Patches

The shape normalized face image is reconstructed by patch-based texture synthesis using patches of the target age. The patch-based texture synthesis can be done by exhaustive search for the patch library at each labeled index (i, j) and tiling selected patches with minimum energy in the raster scan order. The energy function E is defined as weighted sum of the fitness term E_g and the regularization term E_l as shown in Eq. (1).

$$E(i, j, n) = \alpha * E_g(i, j, n) + (1 - \alpha) * E_l(i, j, n) \tag{1}$$

where n represents an unknown patch id in the library, which we would like to find at index (i, j), E_g is the fitness term, which is defined as sum of squared difference between two feature vectors of a shape normalized face image and patches at index (i, j). $\Omega_{i,j}$ represents the region of an indexed patch at (i, j). $\mathbf{I}$ means a feature vector of a normalized face image and $\mathbf{P}$ is that of the patch of the target age. In this paper, a feature vector consists of the vector representation of the RGB pixel intensity $(r, g, b)^T$ and Laplacian of Gaussian filtered response l at pixel (x, y) of a patch, for example, $\mathbf{I} = (r, g, b, l)^t \in \Re^4$ (t is transpose).

$$E_g(i, j, n) = \sum_{x,y \in \Omega_{i,j}} ||\mathbf{I}(x, y) - \mathbf{P}^n_{i,j}(x, y)||^2 \tag{2}$$

E_l is a regularization term, which preserves the spatial coherency between the selected patch and its neighboring patches. In other words, this term preserve visual consistency of the resulting face. Obviously, E_l is calculated by sum of squared difference between feature vectors that both overlap regions between the selected patch and its neighboring patches. In this paper, neighboring patches are on its above $(i, j - 1)$ and left $(i - 1, j)$ locations for (i, j). In our implementation, the squared difference between all possible combinations of patches in each overlap region are calculated and stored at the pre-processing stage. Optimal patches with minimum energy are selected by exhaustive search for the patch library at each index (i, j) in the raster scan order. Of course, overall patch selection result is affected by the processing order depending on the weight for regularization term. Therefore, we need to adjust the weight α so that patches from different persons can be chosen

Fig. 2 Our modified Poisson settings

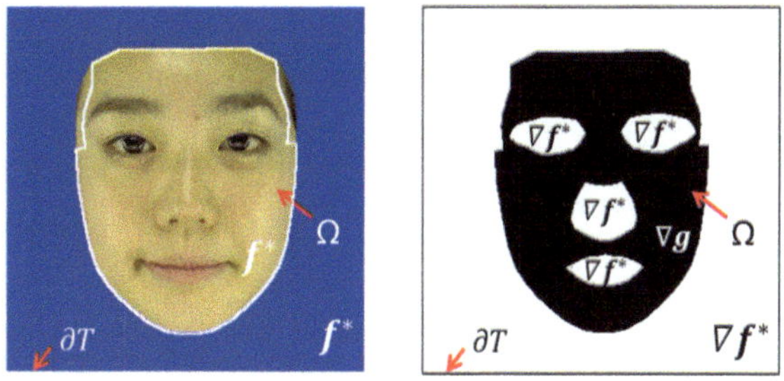

as possible, while preserving visual consistency. We set α to 0.8 empirically (All images in this paper were created by using same α).

Finally, the shape normalized face image is reconstructed by tiling selected patches from top-left to the bottom-right. In this patch-tiling process, the modified Poisson solver [11] seamlessly merges selected patches. Unlike seamless cloning algorithm with naive Poisson [7], the modified Poisson solver can preserve the color of a source image in the composite image by controlling the color preserving parameter. Assuming that **f** is the vector representation of the desired pixel intensity, **r** is also the same representation of the composed image intensity, and **v** is the *guidance* vector field. The modified Poisson problem can be represented by the following equation.

$$\min_f \left(\int_T ||\mathrm{div}\,\mathbf{v} - \Delta \mathbf{f}||_2^2 \, dt + \varepsilon \int_T ||\mathbf{r} - \mathbf{f}||_2^2 \, dt \right) \quad (3)$$

where, T is the region for the whole image, and ε is the color preserving parameter, which preserves the color of a source image in the resulting composite image. If we set a small ε, the resulting image would be affected by the composite image. In our case, we would like to preserve the original skin tone of the input face as much as possible rather than preserving the color of source image. Therefore, we modify the original equation proposed by Tanaka et al. [11] by replacing **r** with $\mathbf{f}^*$ which represents the corresponding intensity of the shape normalized face image (Our setting is depicted as shown in Fig. 2):

$$\min_f \left(\int_T ||\mathrm{div}\,\mathbf{v} - \Delta \mathbf{f}||_2^2 \, dt + \varepsilon \int_T ||\mathbf{f}^* - \mathbf{f}||_2^2 \, dt \right) \quad (4)$$

By discretizing Eq. (4) and vanishing the derivative for a pixel value f_p at a pixel p, we can obtain the following equation.

$$(\varepsilon + |N_p|) f_p - |N_p| \sum_{q \in N_p} f_q = \varepsilon f_p^* - |N_p| \sum_{q \in N_p} v_{pq} \quad (5)$$

where f_p is the pixel intensity at a pixel p and ε is the weight, which decides the effect from an input image. If we set a small ε, the resulting image would be affected by the color of the reconstructed face image. N_p is a set of neighboring pixels at pixel p, and $|N_p|$ is the number of neighbors. Also, v_{pq} is defined by the following equations.

$$v_{pq} = \begin{cases} g_p - g_q & \text{if } p \in \Omega \\ f_p^* - f_q^* & \text{otherwise} \end{cases} \tag{6}$$

where, g_p and g_q represent the pixel intensity at p and q of the reconstructed face image, Ω also represents the region in which the gradients can be transferred (in other words, age-related characteristics can be reflected). We set Ω containing the eyes, nose, and lips to retain the identity of the original face. We consider Eq. (5) for all pixels in the entire image and solve a sparse linear system. Finally, we can obtain the intensity f for each pixel.

4 Statistical Wrinkle Aging Pattern Model

In the texture synthesis phase, occurrence and properties such as shape, depth and length, and the number of wrinkles, pores and pigments of facial skin depend entirely on the pixel intensity pattern in the focusing patch. As mentioned before, it is important to represent fine age-related features such as facial wrinkles, pores, pigments which are strongly affected by age perception [1]. In this paper, we introduce a Statistical Wrinkle Aging Pattern Model (SWAPM) that can implicitly predict the wrinkle appearance, including shape, depth, and length. The SWAPM provides a cue where the appropriate patch can be selected in the patch selection process. The SWAPM can be constructed by carrying out the following procedures using a longitudinal facial image database such as the MORPH database [8]:

(a) Manually marking on left-right laugh lines, left-right crow's feet, and facial wrinkles on left-right orbits and forehead for each image in the database.
(b) Approximating each marked line using a parametric curve and parameterizing each wrinkle by acquired parameters and wrinkle density of each facial wrinkle. More specifically, we use a Ferguson curve for this approximation. The position and velocity at the start/end points of the approximated curve and depth that are decided based on the intensity distribution around the approximated curve are stacked into a vector form for all approximated curves. We refer to the resulting vector as the wrinkle vector for an individual.
(c) Collecting all wrinkle vectors at existing ages in the database and train the aging pattern model in the same manner as the work by Park et al. [5] using principal component analysis. We refer to the resulting linear combination model as the SWAPM.

By changing the coefficient of the SWAPM, we can change the shape, depth, and length of the wrinkles of the model as shown in Fig. 3. The patch selection result can be modulated by adding or subtracting the pixel intensity from resulting wrinkles

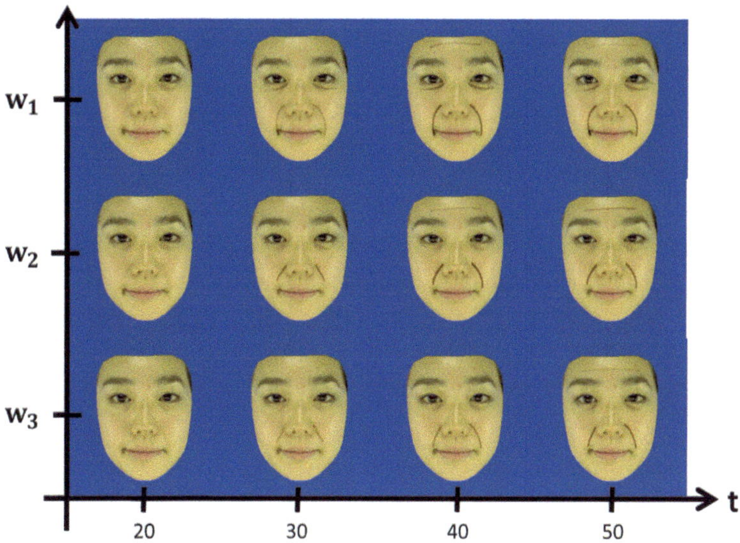

Fig. 3 Changing coefficients of SWAPM. The *vertical* axis corresponds different types of coefficient of SWAPM and the *horizontal* axis means age groups. The *brown lines* represent pseudo facial wrinkles generated by SWAPM. We found that shape, depth, and length of facial wrinkles vary depending on coefficients across the ages

to a shape normalized image. We demonstrate variations of aging simulation for an individual corresponding to Fig. 3 in Fig. 4. We found that our model can describe the variation of aging process for facial wrinkles implicitly.

5 Result and Discussion

Aging simulation results from 10 years old to 60 years for each individual old are shown in Fig. 5. Each row represents aging process for each individual and Each red outlined image represents the original image for each individual. As shown in Fig. 5, our method can represent fine-scale spots and pigments of facial skin that it are difficult to represent using previous methods, as well as age-related facial wrinkles. In addition, unlike previous methods, our method can utilize different image databases for texture synthesis and aging pattern learning. As for texture synthesis, we use facial images of peoples of different ages taken under controlled environment. As for learning SWAPM, we utilize longitudinal facial images for each individual taken under different conditions for facial expression, posture, illumination, and resolution. Thus, we can preserve the quality of resulting texture. In general, it is usually hard or even impractical to collect a large database of large amount of individuals who can provide a series of individual images in different ages under the same shooting condition. Therefore, this point is also advantage of our method.

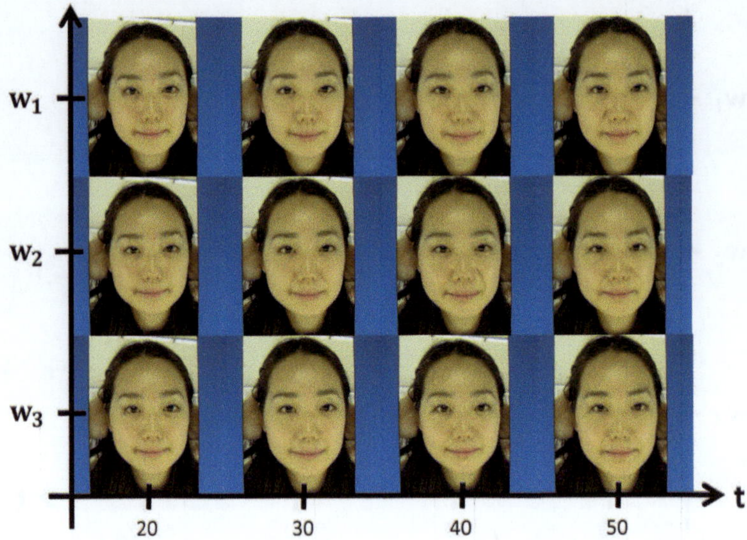

Fig. 4 Result of patch-based texture synthesis for Fig. 3. Similarly, the *vertical* axis corresponds different types of coefficient of SWAPM and the *horizontal* axis means age groups. We observed that variations of wrinkle aging pattern can be represented by SWAPM implicitly

Performance For every synthesized image in Fig. 5, shape normalization and determining wrinkle shape and depth and overwriting pseudo facial wrinkles 0.81 s, patch-based texture synthesis including patch selection and tiling 0.97 s, and geometric deformations 1 s. Entire process takes around 2.78 s in our experiments. Timings are executed on a 2.7 Ghz Intel Core i7 with 8 GB RAM (2011 VAIO Z).

Limitations In patch-based texture synthesis, original features such as moles are sometimes missing or placed in other locations. This is because the gradients of the original image have been completely replaced by those of patches, which consist from others except for the eyes, nose, and mouth region. However, this problem is easy to solve if some user interaction is allowed. More specifically, we manually specify the region Ω in which we would like to retain the identity, in Eq. (6). In addition, we need to consider for optimal size and shape of a patch to improve the quality of synthesized texture (Fig. 5).

Future works The capability to represent other aging effects including increasing/decreasing weight and changes in one's hair is one of the major concerns to be addressed in future work. Moreover, we plan to improve our facial texture synthesis for large pose and illumination changes between an input image and the patch library, and to evaluate the performance of our method on public facial databases like MORPH [8] database by performing a comparison with conventional methods.

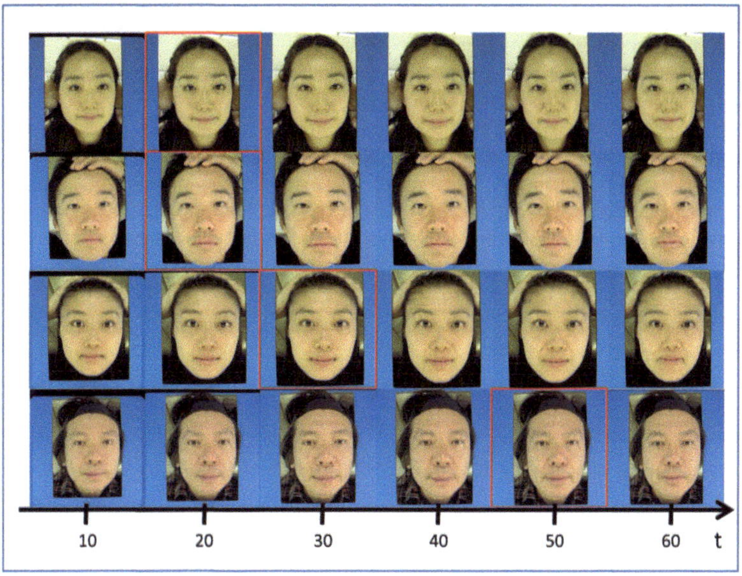

Fig. 5 Aging simulation result from 10 years old to 60 years old. Each *red* outlined image represents the original image for each individual. We observed that our method can represent fine-scale features like spots and pigments of facial skin that it are difficult to represent using previous methods

Acknowledgments This work was supported by the "R&D Program for Implementation of Anti-Crime and Anti-Terrorism Technologies for a Safe and Secure Society", funds for integrated promotion of social system reform and research and development of the Ministry of Education, Culture, Sports, Science and Technology, the Japanese Government.

References

1. Burt M, Perrett DI (1995) Perception of age in adult caucasian male faces: computer graphic manipluation of shape and colour information. J Roy Soc 259:137–143
2. Geng X, Zhou Z-H, Smith-Miles KF (2007) Automatic age estimation based on facial aging patterns. IEEE Trans PAMI 29(12):2234–2240
3. Lanitis A, Taylor CJ, Cootes TF (2002) Toward automatic simulation of aging effects on face images. IEEE Trans PAMI 24(4):442–455
4. Noh J-Y, Fidaleo D, Neumann U (2000) Animated deformations with radial basis functions. In: Proceedings of ACM Symposium on VRST'00, pp 166–174
5. Park U, Tong Y, Jain AK (2010) Age invariant face recognition. IEEE Trans PAMI 32(5): 947–954
6. Patterson E, Ricanek K (2006) Automatic representation of adult aging in facial images. In: Proceedings of the 6th IASTED international conference on VIIP 2006, pp 171–176
7. Pérez P, Gangnet M, Blake A (2003) Poisson image editing. ACM Trans Graph 22(3):313–318
8. Ricanek K Jr, Tesafaye T (2006) MORPH: a longitudinal image database of normal adult age-progression. In: IEEE 7th international conference on FG 2006, pp 341–345

9. Suo J, Chen X, Shan S, Gao W, Dai Q (2012) A concatenational graph evolution aging model. IEEE Trans PAMI 34(11):2083–2093
10. Scherbaum K, Sunkel M, Seidel H-P, Blanz V (2007) Prediction of individual non-linear aging trajectories of faces. Comput Graph Forum 26(3):85–294
11. Tanaka M, Kamio R, Okutomi M (2012) Seamless image cloning by a closed form solution of a modified poisson problem. In: ACM SIGGRAPH ASIA 2012, Posters, 15
12. Tazoe Y, Gohara H, Maejima A, Morishima S (2012) Facial aging simulator considering geometry and patch-tiled texture. In: ACM SIGGRAPH 2012, Posters, 46
13. Zhang L, Tsukiji S, Ai H, Lao S (2005) A fast and robust automatic face alignment system. In: IEEE international conference on ICCV 2005 (Demo Program)

Animating Images of Cooking Using Video Examples and Image Deformation

Syohei Sakiyama, Makoto Okabe and Rikio Onai

Abstract We describe a system that allows users to create animations of cooking from a single picture. Here we focus on images of food being cooked, where steam, bubbles, vibrations of the ingredients, and sizzling sounds are important to convey a depiction of cooking. These elements can be expressed more effectively using an animation than using a single image. However, it may be difficult to record a video of cooking, because the heating involved in cooking changes the fresh and colorful appearance of the ingredients. For these reasons, we propose a method for animating an image of cooking using a combination of bubbles and steam videos, and vibrating ingredients using image-deformation techniques. Although existing video editing software can be used for this purpose, there is typically such a large array of parameters that even professional users may have to invest a significant amount of time to create the animation. Our method semi-automatically determines parameters and allows even novice users to quickly and easily create an animation. Our system allows the user to animate a cooking picture using a few sketch-based inputs in less than 10 min.

Keywords Animating picture · Food · Image deformation · Video database · Video texture · Video segmentation

S. Sakiyama (✉) · R. Onai
The University of Electro-Communications, 2-11, Fujimicho, Chofu, Tokyo 182-0033, Japan
e-mail: sakiyama@onailab.com

R. Onai
e-mail: onai@cs.uec.ac.jp

M. Okabe
The University of Electro-Communications/JST CREST, 2-11, Fujimicho, Chofu, Tokyo 182-0033, Japan
e-mail: m.o@acm.org

K. Anjyo (ed.), *Mathematical Progress in Expressive Image Synthesis I*, Mathematics for Industry 4, DOI: 10.1007/978-4-431-55007-5_21,

Fig. 1 The *left* image shows a picture of a stew. The *right* image is a frame from a video of the stew being cooked

1 Introduction

Images of cooking or dishes are commonly used by restaurants in advertising. Figure 1 (left) shows an example of a stew. Here, we aim to replace such images with animations, because the animation may convey more of an impression of how appealing a dish is by showing rising steam, bubbles, and by the addition of sizzling sounds. However, it may not be desirable to directly record a video of the dish being cooked. Figure 1 (right) shows a frame of a video of a boiling pot of stew; however, it does not appear particularly appealing compared to the picture. The meat and vegetables have changed shape, have sunk down into the fluid, and the overall appearance has lost its vividness. To combine the fresh appearance of the image with the added properties of video images, we propose an interactive method for animating an image. The user animates the picture by interactively specifying the regions where bubbles should appear, as well as the remaining ingredients, and adjusts several parameters describing each region. Then our system automatically creates the animation by making a composite of the original image, video examples of bubbles, and by vibrating the ingredients using image-deformation techniques.

2 Related Works

There are numerous methods for animating images [1, 2]. These are useful for scaling and translating backgrounds, and creating character animations; however, animating images depicting cooking requires specific image-manipulation techniques. Chuang et al. animated an image of rippling water, bobbing boats, swaying trees, and moving clouds using a stochastic motion approach [3]. Furthermore, Okabe et al. created fluid animation using user sketches [4]. Although these methods are useful for providing the textures observed during cooking, they do not allow us to control the parameters describing the appearance of bubbles appropriately.

There are some related methods for synthesizing video textures [5, 6]; however, it is difficult for users to modify the appearance or motion of the resulting animation. Bhat et al. proposed a sketching interface that allows users to edit a video of flowing fluids; however, it does not allow them to animate an image [7]. We require a technique

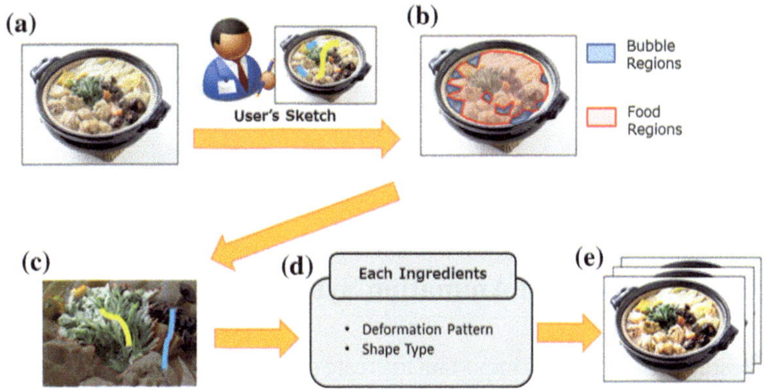

Fig. 2 System overview. **a** Target image. **b** Specify the region of bubbles and foods. **c** Separate the food region into detailed ingredients. **d** Select deformation pattern and shape type. **e** Output

that allows users to create a high-quality animation from a single image of cooking in a short period of time.

3 System Overview

Figure 2a shows the target image; the user begins to create the animation by specifying the regions where bubbles should appear. We use a sketch-based image-segmentation tool [8]. The left mouse button is used to specify the regions where bubbles should appear, and the right mouse button specifies the regions that correspond to ingredients (Fig. 2b). Using the same tool, the user further separates the ingredients into individual items (e.g., pieces of meat and vegetables), as shown in Fig. 2c. For each ingredient, the user selects from one of two parameters: the deformation pattern and the shape type (Fig. 2d).

To apply deformation, the user selects deformable and non-deformable regions. Deformable regions correspond to soft ingredients, such as sliced meats or green leaves. Non-deformable regions describe hard ingredients, such as diced chicken. The user selects one of five shape types: plane, sphere, stick, leaf, or noodle. Given these specifications, our system automatically synthesizes the animation by making a composite video. The parameters include the position, size, and speed of bubbles, and the vibration speed of the ingredients are determined automatically. An animation is then created, as shown in Fig. 2e. Then the user can fine-tune the parameters, if desired.

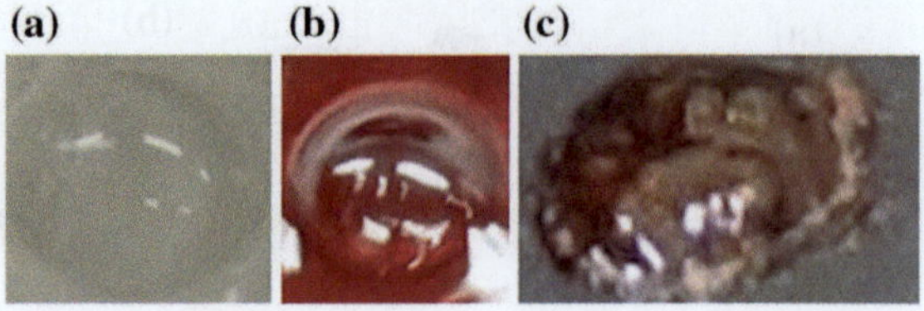

Fig. 3 Our database includes the videos of boiling **a** water, **b** ketchup, and **c** soy sauce

4 Video Examples and Animation Synthesis

The appearance of bubbles is important to create a depiction of cooking. We prepared a database of video examples of bubbles. We recorded several types of bubbles using a video camera. Figure 3 shows three examples of videos of bubbles. We edited each video to prepare multiple versions, by changing the scale and playback speed. These video examples of bubbles are superimposed onto the regions where the user has specified that bubbles should appear. For the example shown in Fig. 2, there are nine regions where bubbles should appear. For each region, i, we consider the area a_i as the number of pixels in the region, and the distance d_i between the center of the region and the center of the dish. We set the size S_i and speed V_i of the bubbles located in the i-th bubble region as follows:

$$S_i = \mu a_i d_i \tag{1}$$

$$V_i = \mu \frac{a_i}{d_i} \tag{2}$$

Then we place bubbles of size S_i to the i-th bubble region on the grid. We apply weightings to the speed of the bubbles. If the bubbles occur along boundaries of the ingredients region, or in a narrow region between ingredients, we apply a factor of 0.7 to reduce the speed of the bubbles. This is based on our observation that the speed of bubbles at boundaries is slower.

We synthesize the animation by making a composite of the input image and the video examples. The synthesis was implemented using the hue saturation value (HSV) color space. We adjust the V channel of the video example so that the brightness matches that of the input image. The process is shown in Fig. 4. For each pixel position (x, y) in frame t o the video example, $VI(x, y, t)$ is the brightness of the input image and $VV(x, y, t)$ is the brightness of the video. We compute the brightness of the synthesized animation $VA(x, y, t)$ using the average of brightness in all frames of the video example, $avg(x, y)$ as follows:

$$VA(x, y, t) = VI(x, y, t) + VV(x, y, t) - avg(x, y), \tag{3}$$

$$avg(x, y) = \frac{\sum_{t=1}^{F} \sum_{x,y} VV(x, y, t)}{N \cdot F}, \tag{4}$$

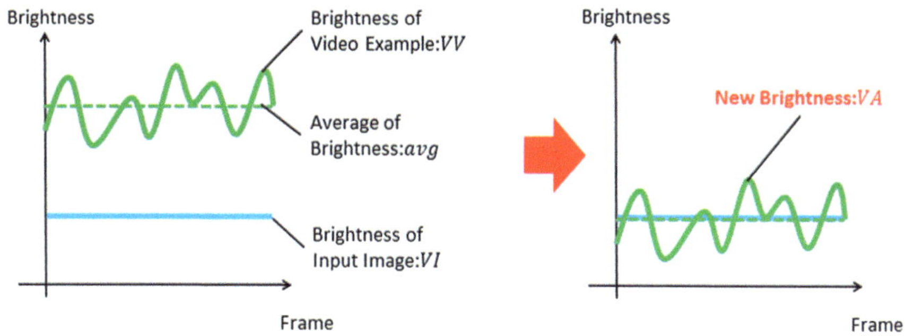

Fig. 4 Adjustment of the brightness in the V channel

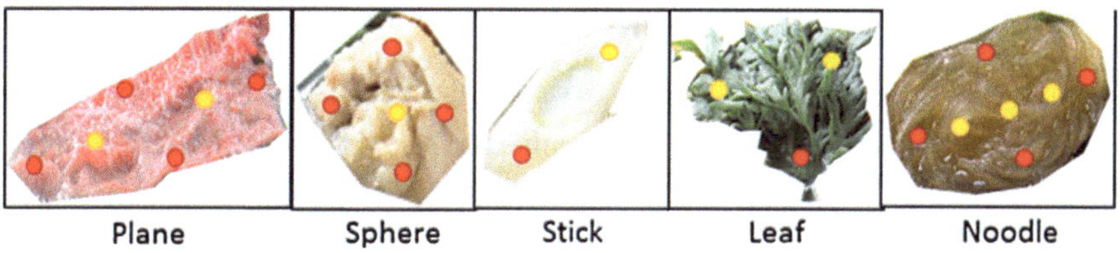

Fig. 5 Placement of control points for each object type. The *red* points are fixed, and the *yellow* points vibrate

where N is the number of pixels and F is the number of video frames. This method enables natural animation synthesis, even if the overall appearance of the video is different from the input image.

5 Vibration of Ingredients

Each of the ingredients is caused to vibrate by the rising bubbles. Our system synthesizes such vibration-mediated animations using a few user-specified parameters. User inputs include the deformation pattern and shape type described in Fig. 2d. Ingredients corresponding to a "deformable" pattern vibrate by deforming the shape. Those with the "non-deformable" pattern vibrate by translation. In deformable patterns, we use rigid transformation using a least squares method [9] to implement the deformation. We create a number of control points, and deform the image by moving these control points. The control points are located according to the shape of the object to be deformed, illustrated by the example shown in Fig. 5. For example, for an image of a slice of meat, which is considered a planar object, the system places six control points, four of which are fixed (shown in red in Fig. 5) and two of which vibrate (shown in yellow). These control points move vertically and horizontally at random within three pixels. Non-deformable objects are not described using control points, and the object vibrates by translating it vertically by three pixels.

Fig. 6 Animations generated using our method **a** 5 min, **b** 8 min **c** 8 min **d** 1 min

6 Results

Figure 6 shows four results of animations generated using our system. We use the water bubbles shown in Fig. 3a in Fig. 6a, b, and use the bubble of the sauce shown in Fig. 3c in Fig. 6c, d. All of these animations were created within a period of a few minutes. To compare the edit times with existing approaches, we synthesized a similar animation to that shown in Fig. 6a using Adobe After Effects (AAE). This required approximately 6 h. The appearance of the animations generated using our approach was slightly less natural than the animations generated using AAE. In particular, the vibrations in our video images looked less natural. However, there is significant scope for improving the algorithms that generate the vibrations and deformations, and our approach allows the user to generate animations considerably more quickly that using existing methods.

References

1. Hornung A, Dekkers E, Kobbelt L (2007) Character animation from 2d pictures and 3d motion data. ACM Trans Graph 26:1–9
2. Igarashi T, Moscovich T, Hughes JF (2005) As rigid as possible shape manipulation. ACM Trans Graph 24:1134–1141
3. Chuang YY, Goldman DB, Zheng KC, Curless B, Salesin DH, Szeliski R (2005) Animating pictures with stochastic motion textures. ACM Trans Graph (TOG) 24:853–860
4. Okabe M, Anjo K, Igarashi T, Seidel H-P (2009) Animating pictures of fluid using video examples. Comput Graph Forum 28:677–686
5. Wei L-Y, Levoy M (2000) Fast texture synthesis using treestructured vector quantization. Proc SIGGRAPH 2000:479–488
6. Bar-Joseph Z, El-Yaniv R, Lischinski D, Werman M (2001) Texture mixing and texture movie synthesis using statistical learning. IEEE Trans Vis Comput Graph 7:120–135
7. Bhat KS, Seitz SM, Hodgins JK, Khosla PK (2004) Flow-based video synthesis and editing. ACM Trans Graph (TOG) 23(3):360–363
8. Kiyono T, Hayashi T, Onai R, Sanjo M, Mori M (2009) Proposal and evaluation of fast image cutout based on improved seeded region growing. Inf Process Soc Jpn 50(12):3233–3249
9. Schaefer S, McPhail T, Warren J (2006) Image deformation using moving least squares. In: Proceedings of SIGGRAPH 2006, pp 533–540

Detection of Inserted Text in Images

Hiromi Hirano, Makoto Okabe and Rikio Onai

Abstract It is possible to embed text into images using an image-editing software, and this can be used to make misleading or unfounded claims in advertisements, which do not comply with advertising standards. To monitor the large volume of advertising that now exists on the Internet, it is desirable to automatically detect and read the text inserted into images. Here we describe a technique to determine regions of images corresponding to inserted text using the FAST algorithm, by finding corners in the image that lie along a straight line, which we term a supercorner. We then create a graph by connecting supercorners and apply cost functions describing the geometrical relations between the corners. Using a graph cut algorithm, we can separate the text from the background. Using this method in a sample set of 130 images with inserted text, we were able to detect 81 % of the inserted text with a false positive rate of only 4 %.

Keywords Supercorner · Fast · Graph cut · Inserted text · Energy function · Edge cost

H. Hirano (✉) · R. Onai
The University of Electro-Communications, 2-11, Fujimicho, Chofu, Tokyo 182-0033, Japan
e-mail: hiromi.hirano@mail.rakuten.com

R. Onai
e-mail: onai@cs.uec.ac.jp

M. Okabe
The University of Electro-Communications/JST CREST, 2-11, Fujimicho, Chofu, Tokyo 182-0033, Japan
e-mail: m.o@acm.org

K. Anjyo (ed.), *Mathematical Progress in Expressive Image Synthesis I*, Mathematics for Industry 4, DOI: 10.1007/978-4-431-55007-5_22,

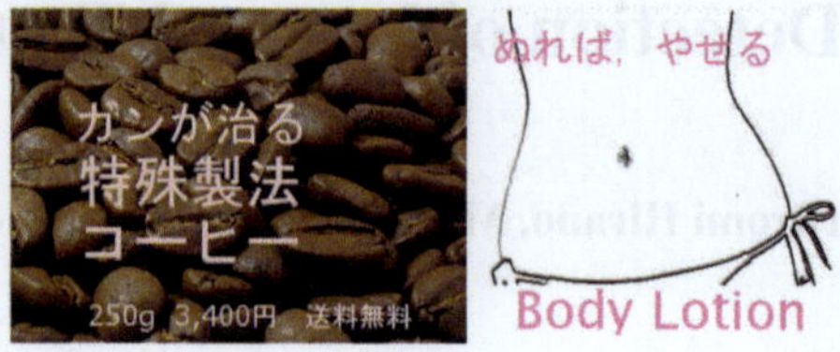

Fig. 1 Illegal image examples; *Left image* says taking this coffee will cure cancer. *Right image* says you can lose weight if you use this body lotion

1 Introduction

A large number of merchant sites advertise their services on the Internet. Some merchant sites present images that do not conform to trading standards regulations, such as advertising goods or services along with unfounded claims. Figure 1 presents some examples, including a bogus weight-loss product (see the image on the right) and claims that do not meet the standards of the Pharmaceutical Affairs Act (see the image on the left). Internet site operators are responsible for detecting and removing such images. Text regions in an image can be classified into two categories: one kind appears on the goods directly, and the other kind is inserted into the image using some kind of image-editing software (see Fig. 3). Here, we propose a method to detect text that has been inserted using image-editing software.

2 Method

Our strategy consists of two stages: first we extract the corners of the image using the FAST [3, 6] algorithm (see Fig. 2); then we determine whether each corner is part of the inserted text region. Corners in inserted text involving Japanese or Kanji characters are likely to fall along a straight line: we refer to this distribution of corners as a supercorner. Additionally, inserted text typically has a large color contrast with the background image, and tends to have a simple color profile. Text that is included as part of the image, such as the labeling on packaging, generally does not fall on a straight line and may have a more complex color profile, making it harder to detect.

We observed that the corners in inserted text exhibited the following trends:

1. the supercorner is located along a line of text;
2. supercorners are typically perpendicular to each other;
3. the corners in a supercorner are of the same color.

We need to assign a label of 'inserted text' or 'background' to each supercorner. To achieve this, a graph cut technique [1] can be used. We construct a graph $\mathscr{G}(\mathscr{V}, \mathscr{E})$, where $\mathscr{V}$ is the set of supercorners, $\mathscr{E}$ is a set of edges between any two supercorners. And each edge has a cost of link (see Sects. 2.3.1 and 2.3.2).

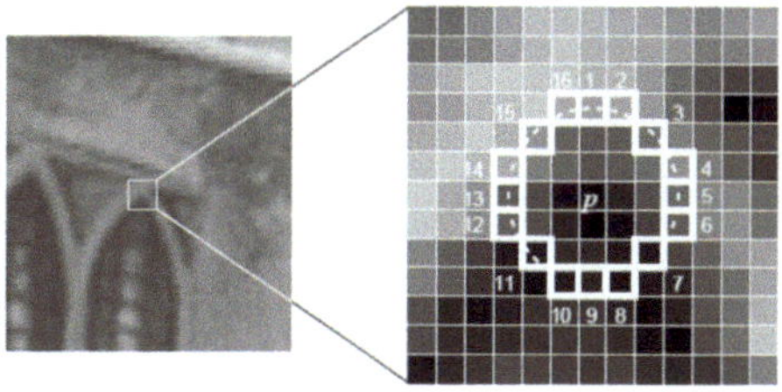

Fig. 2 Corner detection using FAST algorithm. In this chapter, since the radius parameter of FAST algorithm is set to three, we consider the 12 contiguous pixels on the *dashed circle*. When the center pixel p is brighter than all the pixels on the *dashed circle* and the difference of brightness is more than a user-specified threshold, the pixel p is detected as the center of a corner (Referred [3, 6])

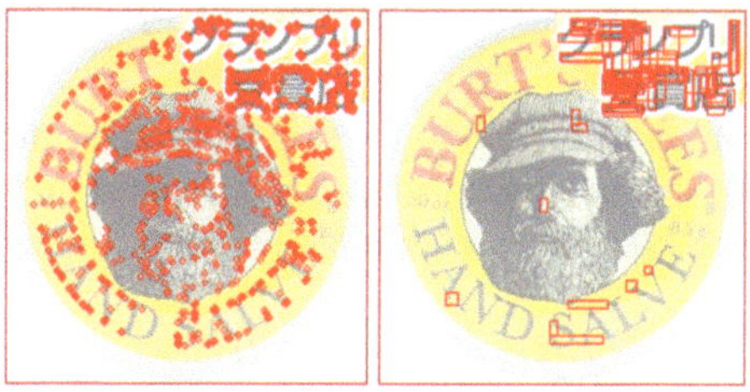

Fig. 3 Extracted FAST and supercorners; *Left* The location of corners in an image extracted using FAST. The *small circle* indicates a FAST corner. *Right* Extracted supercorners

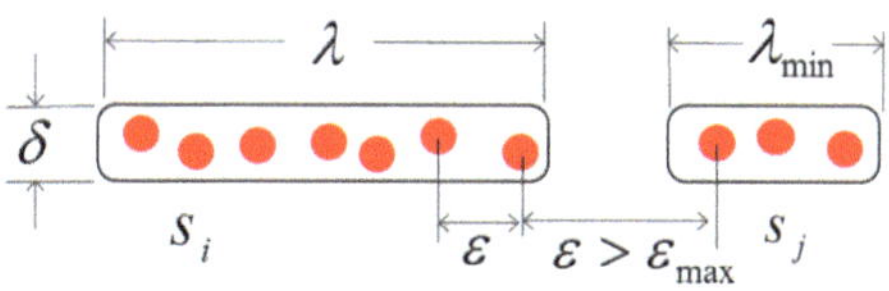

Fig. 4 Supercorner; *Solid circle* shows corner (FAST). We refer to this distribution of corners and bounding box as a supercorner (S_i, S_j)

2.1 The Supercorner

Corners located using FAST may appear anywhere in the image, but corners that lie in a straight line are more likely to form a text region (see Fig. 3). We introduce a line segment with a length that corresponds to the set of corners which we call a supercorner. We start to detect a supercorner by selecting a FAST corner randomly. We then iteratively add the other corners to the supercorner if the following four conditions are satisfied:

1. all FAST corners lie on a line segment within a width δ;
2. the length (λ) of the line is longer than the threshold (λ_{min});
3. the gap (ε) between two FAST corners is shorter than the given threshold (ε_{max});
4. all FAST corners in a set have the same color cluster.

This is illustrated in Fig. 4.

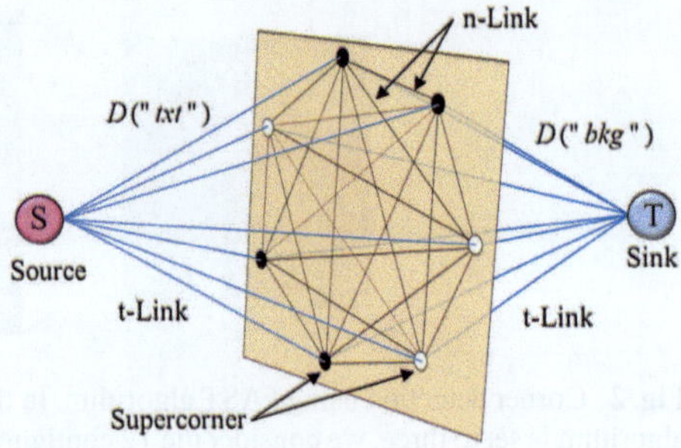

Fig. 5 Supercorner and Graph Structure; *Black* and *white circle* shows a supercorner. *Black* means inserted text area and *white* means others

2.2 Minimizing Energy Function Using Graph Cuts

To apply the graph cut algorithm [5] to the problem, we define an energy function on the graph $\mathcal{G}(\mathcal{V}, \mathcal{E})$. The energy function $E(L)$ is composed of a data term $D(L)$ and a smoothness term $H(L)$. The graph is based on Potts interaction energy model as follows [5]:

$$E(L) = \alpha D(L) + H(L) \tag{1}$$

$$D(L) = \sum_{p \in \mathcal{V}} D_p(l_p) \tag{2}$$

$$H(L) = \sum_{(p,q) \in \mathcal{V}} H_{\{p,q\}} \cdot \delta(l_p, l_q) \tag{3}$$

$$\delta(l_p, l_q) = \begin{cases} 1 & if \ \ l_p \neq l_q \\ 0 & otherwise \end{cases} \tag{4}$$

This energy function can be optimally solved using max-flow [7]. Where L is a label set that indicates whether each supercorner is contained in the text region, so that l_p denotes label of p-th supercorner ($\in \mathcal{V}$), and an α (we choose 1.0) is parameter of the data term. And *label* = {"*txt*", "*bkg*"}.

2.3 Definition of Graph

We construct a graph as shown in Fig. 5, where the n-link edges ($\in \mathcal{E}$) span between two supercorners. The t-link edges ($\in \mathcal{E}$) span between each terminal node and supercorner. A t-link corresponds to a data term and an n-link corresponds to a smoothness term.

2.3.1 Cost of t-Links

We ascribe a cost $\hat{D}_p(l_p)$ to an edge between a supercorner and terminal $\{T, S\}$. The cost is computed using an aligned length λ_p (see Fig. 4) which is normalized to the dimensions of the image size (i.e., the width plus the height). The cost of edges is defined as:

$$\hat{D}_p(l_p) = \begin{cases} D(\text{“}txt\text{”}) = \hat{D}(\lambda_p) \\ D(\text{“}bkg\text{”}) = 1 - \hat{D}(\lambda_p) \end{cases} \tag{5}$$

where the labels “txt” and “bkg” denote text and background regions of the image, respectively. And $\hat{D}(\lambda_p)$ is defined as a monotonically increasing function whose value range is from 0 to 1, $\sigma^2_{\lambda_p}$ is a variance. See Eq. (6).

$$\hat{D}(\lambda_p) = 1 - \exp\left(-\frac{\lambda_p^2}{2\sigma^2_{\lambda_p}}\right), \ \lambda_p \geq 0 \tag{6}$$

2.3.2 Cost of n-Links

$H_{\{p,q\}}$ in Eq. (3) represents a cost between node p and q. As we refer to a supercorner as a node, we can give $p \equiv s_i$, $q \equiv s_j$, and Eq. (7).

$$H_{\{p,q\}} \equiv H(s_i, s_j) = A(s_i, s_j) + B(s_i, s_j) + C(s_i, s_j) \tag{7}$$

where $H(s_i, s_j)$ is defined as a summation of following three costs.

1. If two supercorners are parallel and in close proximity, then the edge cost is low. The function $eval(s_i, s_j)$ is minimized when two supercorners are in close proximity and parallel to each other as follows:

$$A(s_i, s_j) = 1 - \exp\left(-\frac{eval(s_i, s_j)}{2\sigma^2_{eval}}\right) \tag{8}$$

2. If two supercorners are (nearly) orthogonal, then the edge cost is low. The function $orth(s_i, s_j)$ reflects mutually orthogonal relation between s_i and s_j, as follows:

$$B(s_i, s_j) = 1 - \exp\left(-\frac{orth(s_i, s_j)}{2\sigma^2_{orth}}\right) \tag{9}$$

3. If two supercorners are determined to be in same color cluster, then the edge cost is low. The function $cdist(s_i, s_j)$ describes the color distance between s_i and s_j, as follows:

Table 1 Performance comparison between two methods on same data set

Method	N_a	N_b	N_d1	N_d2	Precision (N_d2/N_d1)	Recall (N_d2/N_b)
Our method	135	78	92	75	0.815	0.962
Zhao et al. [8]	135	78	194	65	0.351	0.833

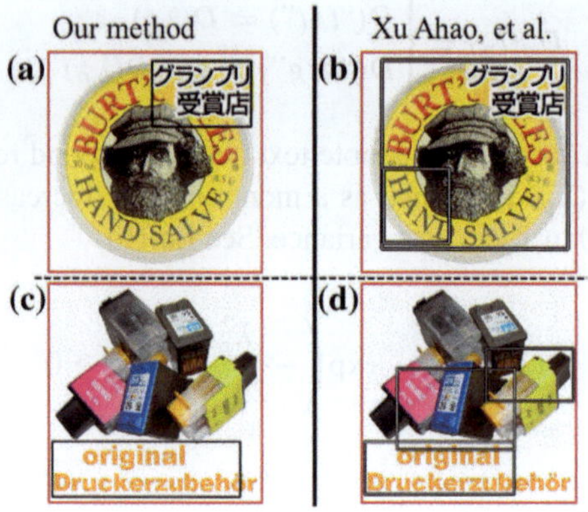

Fig. 6 Comparative evaluation; Our method correctly extracts an overwritten text area. See (**a**) and (**c**). In contrast, the comparative approach generates wrong areas. See (**b**) and (**d**)

$$C(s_i, s_j) = 1 - \exp\left(-\frac{cdist(s_i, s_j)}{2\sigma_{cdist}^2}\right) \tag{10}$$

In Eqs. (8)–(10), σ_*^2 represents a variance.

3 Comparative Evaluation and Result

We conducted a comparative evaluation using an existing proposed method by Zhao et al. [8]. They use distribution characteristic of corners extracted from a gray scale image by Harris corner detector. Such corners may appear at anywhere in images. It cannot distinguish an inserted text from a text that is included as part of an image, such as labels on a package. Our approach uses linearly arranged FAST corners and color at corner, then generates a graph which has supercorners as nodes and relations between any two supercorners as edge. And each edge has a weight that is computed using equation from (7) to (10). And we apply graph cut algorithm to this graph. Then our approach can distinguish them.

We applied both methods to 130 images sampled from Rakuten Mart. Table 1 shows the result. Where N_a is a number of text, N_b is what N_a minus a number of printed text on a surface of goods. N_d1 is a number which each method detects as

Fig. 7 Some results; All images show that it correctly detects inserted text area

text area. N_d2 is a number of correct cases in N_d1. So we compute a precision = N_d2/N_d1 and a recall = N_d2/N_b.

Our method identified 81 % of inserted text with a false positive rate of only 4 %. Figure 6 shows some inserted text area which is detected by each method. Our method correctly detect the inserted text in images. Figure 7 provides some sample images with the inserted text identified. Therefore, the algorithm is effective as a filter tool to detect inserted text in images.

Finally, we will make a point that all images described here without Fig. 1 are not illegal.

References

1. Boykov Y, Veksler O, Zabih R (2001) Fast approximate energy minimization via graph cuts. In: IEEE Transactions on PAMI, pp 1222–1239
2. Gopalan C, Mamjula D (2008) Contourlet based approach for text identification and extraction from heterogeneous textual images. Int J Comput Sci Eng 2(4):202–211
3. Rosten E, Drummond T (2005) Fusing points and lines for high performance tracking. In: 10th IEEE international conference on computer vision, 2005. ICCV 2005, vol 2, pp 1508–1511
4. Gatos B, Pratikakis I, Kepene K, Perantonis SJ (2005) Text detection in indoor/outdoor scene images. IEEE Trans Pattern Anal Mach Intell 127–132
5. Kolmogorov V, Zabih R (2004) What energy functions can be minimized via graph cuts? In: 1st international workshop on camera-based document analysis and recognition, vol 26, no 2, pp 147–159
6. Rosten E, Drummond T (2006) Machine learning for high-speed corner detection. In: European conference on computer vision, pp 430–443
7. Wayne K (2004) Max flow, min cut. In: Algorithms and data structures, Princeton University, COS 226
8. Zhao X, Lin K-H, Liu Y, Huang TS (2011) Text from corners: a novel approach to detect text and caption in videos. IEEE Trans Image Process 20(3):790–799

Index

K. Anjyo (ed.), *Mathematical Progress in Expressive Image Synthesis I*,
Mathematics for Industry 4, DOI: 10.1007/978-4-431-55007-5,

MIX
Papier aus verantwortungsvollen Quellen
Paper from responsible sources
FSC® C105338

If you have any concerns about our products,
you can contact us on
ProductSafety@springernature.com

In case Publisher is established outside the EU,
the EU authorized representative is:
Springer Nature Customer Service Center GmbH
Europaplatz 3, 69115 Heidelberg, Germany

Printed by Libri Plureos GmbH
in Hamburg, Germany